Praise for Sarah Wise

Inconvenient People

SHORTLISTED FOR THE WELLCOME BOOK PRIZE

'Deeply researched and gripping . . . The book owes its enormous power to Sarah Wise's patience. She has sifted through hundreds of case histories . . . It makes for harrowing reading, but much of it is also hilarious.'

A.N. Wilson, *Mail on Sunday*

'This gripping study of the Victorian madhouse . . . sends some shivers up the reader's spine along the way.'

Wall Street Journal

'Wise has a wonderful series of gothic tales to tell . . . She is the first to draw them together into a broader portrait of the fraught relations between Victorian mad-doctors and at least some of their patients. Moreover, she has worked diligently in the archives to uncover fresh details that make the curious characters that populate her pages come to life.'

Times Literary Supplement

'After these cheerful late cases comes a devastating epilogue . . . you put this quite superlative book down, shaken.'

Independent

'Sarah Wise is an excellent writer, and those who pick up this book will not lightly put it down. Her ten chapters read like short novels, and she has the true social historian's ability to make her period come alive. She selects and compresses the salient details beautifully; one often feels as if one is actually present at the scenes she describes. There can be no higher praise.'

Spectator

'Fascinating . . . Sarah Wise has used her subject like an axe, to split open the Victorian façade and examine everything wriggling behind. It has enough tragedy, comedy, farce and horror to fill a dozen fat novels, and enough bizarre characters to people them.'

Financial Times

'Fascinating and chilling, *Inconvenient People* reads like a series of Victorian novels in brief – only all the tales are true. Each chapter is like a complex costume drama, involving plots, firearms, escapes, pursuits, confrontations, angry crowds, abandonment, terror and despair.'

Daily Mail

The Blackest Streets

Shortlisted for the Royal Society of Literature's Ondaatje Prize

'There are those writers whose imagination lives and prospers within a certain territory; for them the ground itself is haunted, a terrain upon which the past and present are locked in an embrace like lovers. So it is for Sarah Wise . . . An excellent and intelligent investigation of the realities of urban living . . . This is a book about the nature of London itself.'

Peter Ackroyd, *The Times*

'Scrupulously researched and eye-opening . . . a revelatory book, tearing the roofs off the Old Nichol's festering tenements, beaming the light of impartial historical research into the horrible dens and alleys, exposing the blighted lives and the crushing deprivation.'

Professor John Carey, *Sunday Times*

'Remarkable . . . Wise is a rigorous historian, but it is her subtle ability to summon the individual to elucidate the whole without ever resorting to stereotype or simplification that makes this book so memorable . . . This engrossing work shines a light not only on a turbulent period in London's history but on humanity itself. Only the best histories can claim as much.'

Guardian

The Italian Boy

WINNER OF THE CRIME WRITERS' ASSOCIATION
GOLD DAGGER FOR NON-FICTION
SHORTLISTED FOR THE SAMUEL JOHNSON PRIZE FOR NON-FICTION

'I wish I had written *The Italian Boy*, but alas, Sarah Wise got there first, and did so in style . . . This is an impressive debut and a compelling piece of history writing.'

Bernard Cornwell, *Mail on Sunday* Books of the Year

'A haunting blend of scholarship and period empathy.'

Iain Sinclair, *Daily Telegraph*, Books of the Year

'An amazing book . . . It out-Dickenses Dickens.'

Dr Maria Misra, Samuel Johnson Prize judge

Also by Sarah Wise

*Inconvenient People: Lunacy, Liberty
and the Mad-Doctors in Victorian England*

The Blackest Streets: The Life and Death of a Victorian Slum

The Italian Boy: Murder and Grave-Robbery in 1830s London

The Undesirables

The Law that Locked Away a Generation

Sarah Wise

ONEWORLD

A Oneworld Book

First published in Great Britain, the Republic of Ireland and
Australia by Oneworld Publications, 2024

Copyright © Sarah Wise, 2024

The moral right of Sarah Wise to be identified as the Author of this work has been
asserted by her in accordance with the Copyright, Designs, and Patents Act 1988

All rights reserved
Copyright under Berne Convention
A CIP record for this title is available from the British Library

ISBN 978-0-86154-455-4
eISBN 978-0-86154-456-1

Typeset by Hewer Text UK Ltd, Edinburgh
Printed and bound in Great Britain by Clays Ltd, Elcograf S.p.A.

Oneworld Publications
10 Bloomsbury Street
London WC1B 3SR
England

Stay up to date with the latest books,
special offers, and exclusive content from
Oneworld with our newsletter

Sign up on our website
oneworld-publications.com

In memory of my uncle, Ian Wise, 1931–2020, a liberated captive.

Contents

Unkind Words: A Note on Terminology	xi
A Note on Anonymity	xiii
Foreword	xvii

Part 1: Inevitability

1. Winning the Argument	3
2. Occasional Criminals and Natural Monsters	10
3. 'Education on an Empty Stomach'	19
4. 'Human Dregs at the Bottom of Our National Vats': The Collapse in the Middle-Class Birth Rate	31
5. 'One Law for Men, and Another Law for Women'	35

Part 2: Changing the Law

6. 'Be Very Careful': Winston Wades In	55
7. The Man Who Spoke Drivel	70

Part 3: 'Defects of Character and Temperament'

8. 'A Cross Between Public Schools and Guild Training Colleges': The Mental Deficiency Colony	89
9. 'She Has a Bad Name in Clifton Hampden': Bessie B— and the Policing of Female Sexuality	107
10. 'A Crooked Mind': The Adolescent Thief	127
11. Dirty Homes, Stupid Kids	140
12. Lovesick in St James's	156
13. 'And They Catch 'im, and They Say 'e's Mental': The Sex Offenders	167

Part 4: Solutions

14. The Gateshead Castrations, or Paganism and the Knife	183
15. Unnatural Selection: What Went On Abroad	198
16. The East Chaldon Vicarage Affair: The 'Feeble-Minded' in the Community	210

Part 5: Liberation?

17. 'A Scandal on a Far from Small Scale'	229
18. The Keys to the Door	242
19. Forty-Six Years of Waste	256

Afterword: The Way We Live Now – The Deprivation of Liberty	259
Appendix 1: How the Mental Deficiency Act Worked	265
Appendix 2: Anecdotes of Women Detained for Having a Child Out of Wedlock	267
Appendix 3: American Responses to a British 1951 Survey on Sterilisation	272
Acknowledgements	277
Notes	279
Bibliography	303
Picture Credits	313
Index	315

Unkind Words: A Note on Terminology

In this book, the insensitive terms 'feeble-minded', 'weak-minded', 'mentally deficient' and 'mentally defective' reflect the dominant attitudes of the period under consideration. These are historically accurate words for today's more acceptable 'learning-disabled'. The use of such language does not imply the author's acceptance of these categories.

'Mentally defective' and 'mentally deficient' are used interchangeably.

In 1927, the term 'moral defective' replaced 'moral imbecile' – the original 1913 diagnostic category.

Below is the categorisation used during the framing of the 1913 Mental Deficiency Act. *The Undesirables* concerns itself solely with diagnostic categories 5 and 6:

1. Persons of unsound mind, roughly equivalent to the term lunatic: persons who require care and control owing to disorder of mind and consequently incapable of managing their affairs.

2. Persons mentally infirm, who through mental infirmity, arising from age or from decay of their faculties, are incapable of managing themselves or their affairs.

3. Idiots: persons so deeply defective in mind from birth or from early age that they are unable to guard themselves from common physical dangers such as, in the case of young children, would prevent their parents from leaving them alone.

4. Imbeciles: persons who are capable of guarding themselves against common physical dangers, but who are incapable of earning their living by reason of mental defect existing from birth or early age.

5. Feeble-minded: persons who may be capable of earning a living under favourable circumstances but are incapable, from mental defect existing from birth or early age, of a) competing on equal terms with their normal fellows, or b) managing themselves and their affairs with ordinary prudence.

6. Moral imbeciles: persons who from an early age display some mental defect coupled with strange, vicious or criminal propensities on which punishment has little or no deterrent effect.

From the start, the porous nature of these classifications would cause problems, and even many specialists in the field were muddled on the precise definition of 'feeble-minded'. The most helpful explanation (for us) comes from doctor Sir James Crichton-Browne, who stated in 1905:

A feeble-minded person is one who by reason of arrested development or disease of the brain dating from birth or from some age short of maturity has his observing and reasoning faculties partially weakened, so that he is slow or unsteady in his mental operations, and falls short of ordinary standards of prudence, independence and self-control. It is a very difficult thing to define, especially as we have not got a definition of what normal is.

A Note on Anonymity

In these pages, I have disguised the names of individuals that have not so far appeared in the public record, in order to respect the sensitivities of relatives and descendants.

All other names are real, having already appeared in newspapers, magazines or other public documents.

Q: 'Could you put in words a definition of feeble-minded, so that people could test the condition of any person who was to be subjected to the operation of the law?'

A: 'No, I could not.'

Q: 'Have you any suggestion to make as to how we could get an adequate definition for the purpose of a statute or rule?'

A: 'No, I rather doubt it being possible.'

Answers of Herbert Jenner-Fust, Local Government Board inspector, questioned by the Royal Commission on the Care and Control of the Feeble-Minded, 1908

'Wherein lies their defect? It is not defect of knowledge, of intelligence in the educational sense, but failure to adjust themselves to the social conditions in which they live.'

Dr Rees Thomas, medical superintendent of Rampton State Institution, 1920

'A detention as a moral defective is often a detention for life; it is a detention, pure and simple, without treatment and without limit . . . In fact, it is the writing off of a life at the age of eighteen as worthless.'

The National Council for Civil Liberties, '50,000 Outside the Law', report, 1951

Foreword

In the late 1940s, the National Council for Civil Liberties announced that around fifty thousand young people were being detained in institutions on the grounds that they were 'mentally deficient'. The youngsters had been given no psychiatric diagnosis, they were offered no treatment, most had committed no crime, and there was no time limit on their detention. The NCCL stated: 'The cases of detention under the mental deficiency laws will come as a shock to all. There is, amongst the general public, little knowledge of the Mental Deficiency Service. It constitutes one of the gravest social scandals of the twentieth century. It is almost unbelievable that such cases can occur in England today.' Thirty per cent of the people certified as mentally deficient had been in detention for between ten and twenty years – five per cent for over thirty years.

The Undesirables is my account of how such a state of affairs came about and an attempt to piece together fragments of the experiences of those who underwent such detention.

In the opening chapters that follow, I'll explore a number of 'panics' that intertwined and created the terrain in which the Mental Deficiency Act took root. The 1913 Act permitted local authorities to identify children and young adults they believed incapable of good behaviour – typically, young hooligans and persistent thieves; girls who had given birth to an illegitimate child; and anyone who seemed either incapable of or uninterested in working, earning and living independently. The council mental deficiency executive officer was known in some localities as the 'rat-catcher'. However, an uncomfortable truth is that parents and relatives were often the prime movers in having a difficult or challenging youngster placed in a institution.

This is not a book about mental disability/learning difficulty: instead, it concerns individuals detained for social or 'moral' reasons under the new 'moral imbecile' category, introduced in the 1913 Act. This was a diagnosis by which children and adolescents who appeared to be showing anti-social tendencies could be sent to sex-segregated 'colonies' for life. This was so they could not breed the next generation of undesirable Britons. As high-profile eugenicist Alfred Tredgold put it: 'Their propagation must be prevented.' It was the closest this country came to a selective breeding campaign. This form of preventive detention – for life – had no precedent in England. *The Undesirables* traces how the Mental Deficiency Act came to be passed on the eve of the First World War, and how such an attack upon personal freedom won parliamentary approval – and how the eugenicist fears of 'degeneration' gained the upper hand. England had, across just over a decade, moved from an idealisation of personal liberty to the polar opposite – granting the state the power to detain someone for life on the supposition of what they might do in future, rather than for an offence that had been proven in court.

It was unclear whether moral imbeciles were simply people who lacked an inherent moral sense (if such a thing existed, which was heatedly contested), or whether moral inadequacy had to be coupled with an underlying mental defect. Could 'incomplete emotional development' correctly be described as a biological problem? It was perplexing that perhaps the majority of those put away as 'moral imbeciles' showed little clear evidence of learning impairment, and thus the diagnosis tended to be a personal judgment on the part of a doctor. In fact, in 1950 two psychologists random-sampled one hundred patients at a large mental deficiency hospital and found that one-quarter had an IQ of seventy or above (seventy being the accepted borderline between 'normal' and 'subnormal' intelligence).

People deemed to be 'socially inefficient' were highly vulnerable to being declared defective, unable to take their proper place in a nation that required a skilled workforce – a likely burden to Britain in its bid to retain pre-eminence in an internationally competitive world. They were seen as a brake, and an expensive one too – as they littered up workhouses, prisons and reform schools; and they were also assumed to be holding back their own families and communities.

The Mental Deficiency Act continues to reverberate into our own age. Many people have in their family tree forebears who spent years locked away. The sense of shame this can cause can still be keenly felt by descendants. When the great age of release arrived, from the 1959 Mental Health Act onwards, it was often assumed that this was a nineteenth-century barbarism that was being corrected – that it was instituted by a famously repressed and hypocritical Victorian culture. *The Undesirables* hopes to lay that particular myth to rest.

PART 1

INEVITABILITY

1

WINNING THE ARGUMENT

When Charles Darwin's cousin Francis Galton introduced the concept of eugenics in 1883, many influential commentators declared that poor educational attainment, as well as anti-social behaviour and criminality, were transmitted down the generations, and were rarely the result of environmental, societal or cultural factors. While methodologically rigorous statistical surveys to prove this point would never emerge, the rhetoric of the 'social Darwinists' was powerful – emotive, portentous and persuasive.

> Even when their houses are whitewashed, the sky will be dark; devoid of joy, they will still tend to drink for excitement; they will go on deteriorating; and, as to their children, the more of them grow up to manhood, the lower will be the average physique and the average morality of the English generation.

Economist Alfred Marshall wrote these words in the prestigious journal *The Contemporary Review*; and prominent eugenicist George Mudge would go even further, stating that

> the environment is the product of the individual, and not vice versa. The stunted individuals are not the product of a one-roomed tenement, but the one-roomed tenement is the expression of the inherent incapacity of this race to be able to do anything better for itself . . . it is the natural outcome of their already existing physical, moral or intellectual degeneration. These degenerates are 'mutations', and breed true to their degeneracy.

Such views were always forcefully challenged, and even at the height of their influence, hereditarians would never win unequivocal backing. Novelist and columnist Clarence Rook, in his article 'St Patrick Hooligan', on London street crime, published in 1900, believed no innate difference existed between boys growing up in chronic poverty and those who had been to public school and university:

> So long as we leave several hundreds of thousands of boys to roam the streets with no legitimate outlet for their abundant energy, so long shall we be startled by the howl and occasionally stunned by the belt of the Hooligan ... Let us remember that the Oxford undergraduate was a Hooligan, delighting in town-and-gown rows, until someone had the happy thought of turning his misdirected energy towards athletics.

Helen Bosanquet, meanwhile, asserted that urban conditions were the source of the seeming 'degeneration' of Londoners, writing sceptically in 1895:

> The excitement of a town life tells very greatly upon children; if you look closely, you will see that London children are always tired; the dark rings under their eyes tell of the nervous strain which is breaking down their health, and their very restlessness is the restlessness of fatigue and nervous exhaustion ... [They] are generally in the third generation of London life. But to say this alone is at once too much and too little; it implies a cumulative and inevitable evil in which I do not believe, and the fatalism of the observation seems to yield a little before analysis.

Dig a little deeper, think more carefully, said Bosanquet, and hereditarian views do not stand up. But environmentalist arguments such as these lacked the vigour and confidence of hereditarian prose. The drift towards fatalism continued, culminating in the remarkably draconian 1913 act of parliament, the Mental Deficiency Act, to deal with the supposed ongoing deterioration of the British mind and morals.

Five factors fuelled this over-reaction:

> the study of prisoners (particularly recidivists) by medical officers in gaols;

the findings of school authorities, noting the unexpectedly large number of children who struggled to acquire even the most basic skills;

the fall in the birth rate among the middle and upper classes, and the apparent super-fertility of the poor – with the allegedly 'feeble-minded' poor cited as especially fecund;

'rescue societies' data on girls and women with illegitimate children, or who appeared to be particularly vulnerable to sexual assault or sexual coercion;

and the pessimism of many involved in mental healthcare about the curability or treatability of certain psychological conditions.

***

To take the last first: heredity's role in mental health problems was far more frequently discussed after publication of Darwin's *Origin of Species* in 1859 and subsequent works by popularisers of his particular theory of evolution. However, pre-Darwinian ideas about the inheritance of psychological traits frequently crop up in the writings of eighteenth- and early nineteenth-century 'doctors of psychological medicine'.

In the 1820s and 1830s some doctors focused on the potential problems in store for the aristocracy and the upper-middle classes; 'cousin marriage', in particular, was blamed for an observable rise in offspring who were 'idiotic', or 'imbecilic', or 'insane'. George Mann Burrows, the trusted doctor of many wealthy families, wrote in 1828 that

> hereditary predisposition is a prominent cause of mental derangement... Among the highest ranks, hereditary insanity is more common than among the lower; for the former most frequently contract marriages with their own rank, or even their own family. Hence, wherever the system of clanship or family connexion has been most strictly preserved, there it most prevails.

Burrows, and other specialists who sought the banning of marriage between cousins, had solely this kind of anecdotal evidence to offer, as no meaningful statistical analysis was possible at that point. The emphasis on the offspring of the wealthy is worth noting; at around the middle of the nineteenth century, the focus of alarm would shift to the poor.

A near-contemporary of Burrows, James Cowles Prichard, was an influential evolution theorist and a doctor who had studied the patients of large lunatic asylums in both Britain and France. In his 1835 *Treatise on Insanity and Other Disorders Affecting the Mind*, Prichard suspected that there may be a hereditary factor in a condition he termed 'moral insanity'. Prichard wondered if a child could inherit a faulty 'organisation' within the nervous system that caused an imbalance between the 'passion' and the 'will'. Although he had not devised this disease category (its roots lay in France in the eighteenth century), Prichard developed the theme for the Anglophone world. Moral insanity posited a type of mental disease in which there is no delusion, and intellectual and analytical faculties are unaffected; but the sufferer's moral and social sense are severely abnormal, affecting their ability to live in the world, and causing distress and even danger to those among whom the sufferer lives.

Prichard wrote that moral insanity

> is a morbid perversion of the natural feelings, affections, inclinations, temper, habits, moral dispositions and natural impulses, without any remarkable disorder or defect of the intellect, or knowing and reasoning faculties, and particularly without any insane delusion or hallucination . . . The subject is found to be incapable, not of talking or of reasoning . . . for this he will often do with great shrewdness and volubility, but of conducting himself with decency and propriety in the business of life.

Moral insanity proved to be a hugely controversial theory in Britain and triggered many legal and medical battles across the nineteenth century as the courts tested the borderline between 'normal' eccentricity and abnormal, perverse behaviour indicating mental derangement. These arguments took place during formal lunacy hearings to decide an alleged lunatic's true state of mind, as well as during criminal trials to judge the sanity (and therefore culpability) of defendants being prosecuted for violent crimes.

In 1857 French doctor Bénédict Augustin Morel published his influential *Traité des Dégénérescences Physiques, Intellectuelles et Morales de l'Espèce Humaine* (Treatise on the Intellectual, Moral and Physical Degeneration of the Human Race), in which he proposed a model by which inherited traits may lead to the ultimate annihilation of a family line. In his model, an immoral act committed consciously by someone 'normal' would produce 'nervousness' in the next generation, insanity in the third generation, idiocy in the fourth, and finally extinction. Morel's mental degeneration process was accompanied by observable physical differences, for example in head shape and size, ear and eye-pupil size, and changes to various reflexes. Morel believed that the primary event (the original immoral act) could happen to anyone, but that the course of the subsequent degeneration in their offspring could be predicted by a well-trained doctor.

Morel's seeming breakthrough occurred as asylum population levels were rising (which most commentators attributed to an increase in mental illness, rather than to greater reporting and increasing numbers of people being brought forward for treatment); but also at a time when it looked as though the limits of psychological medicine had been reached. Morel's work offered an explanation for the incurability of a wide range of patients: a large proportion of these disordered minds and nervous systems, it now seemed clear, were untreatable – not through any fault of doctors, but because these individuals sat somewhere on Morel's matrix of degeneration.

Many influential British 'alienists' (psychiatrists) of the 1860s to 1890s, including Daniel Hack Tuke and Henry Maudsley, also believed that the moral function was linked to familial inheritance. In Tuke's theory of 'dissolution', the evolution of the human nervous system had developed the 'higher' faculties, such as self-control, a social sense and a sense of propriety; and reduced such 'lower' instincts as selfishness, greed, sexual promiscuity, slothfulness and alcoholism. In moral insanity, the higher functions were the first to become diseased, Tuke wrote, allowing crude, atavistic instincts to dominate the patient's behaviour. In some way not yet understood, this marked a regression to an earlier state of humankind, caused by the inheritance of a weakly constituted mind and nervous system.

For Henry Maudsley, the 'struggle for life' had become increasingly intense in the so-called 'civilised' nations. He wrote that this struggle created anxiety, which impacted adversely upon the most weakly constituted minds – minds of inferior human stock whom he described as 'abortive beings in nature' and 'social wrecks'. Maudsley thought Herbert Spencer's 1864 term 'the survival of the fittest' to be 'the most felicitous phrase of our epoch . . . It applies as much to the world of morals as to the world of the intellect and feeling and action.' Maudsley had been hugely influenced by Morel's work and came to believe that in many sufferers, 'moral insanity' was an illness that arose from a congenital defect – that is, an inherited mental weakness from birth that predisposed an individual to developing that particular form of mental illness in adult life.

Why did this line of thinking come to such prominence in these years? As noted, degeneration suggested an answer to the distressing fact of the intractability of many mental conditions. As highly experienced doctors, Maudsley and Tuke had years of coming to terms with the failure of psychiatry to alleviate human suffering; and the idea of the biological inevitability of weak or diseased minds made sense of a harrowing and frustrating truth.

But there's more to it. Britain was an increasingly competitive society: roles and stations in life previously off limits to the ordinary man were now, slowly, opening up to those with talent and energy. (And Britain itself was facing an increasingly competitive world of globalising trade.) These were the 'struggles' Maudsley believed the 'civilised' were facing. By contrast, he assumed that 'savages' lived comparatively mentally healthy lives, unmolested by ambition, over-work and anxiety about social status.

Francis Galton devised the neologism 'eugenics' (in preference to his original choice, 'viriculture') as part of his investigation into 'the conditions under which men of high type are produced'. His men of high type comprised the geniuses, of course; but they also included successful members of the professional classes, whose superior qualities were demonstrated by the very fact that they had achieved social standing and a degree of wealth. The concepts of 'civic worth' and 'social value' came to prominence in the 1880s, and Galton produced many pedigree charts to prove that eminence was transmitted biologically. It was the middle-class man (ranging from the highly educated professional to the skilled petit bourgeois artisan) in whom superior stock was to be found. Men such as Galton, Maudsley and Tuke.

This was a new democratic form of elitism – a revolution, really: position in life would henceforth be achieved because of inherited intellectual and moral superiority, rather than inherited wealth or titles. An effete and enervated aristocracy and landed class, its stock weakened by centuries of inbreeding, would be supplanted when the naturally endowed began to select each other to marry and produce superior offspring. 'What Nature does blindly, slowly and ruthlessly, man may do providently, quickly and kindly', Galton wrote, grossly overestimating the speed at which human evolution occurred.

It wasn't just the mad doctors who felt inspired by the concept of degeneration. The model also offered an explanation for a range of seemingly intractable, incorrigible criminal and anti-social behaviour, too. So could 'social failure', and even chronic poverty, also be attributed to degeneration?

2

OCCASIONAL CRIMINALS AND NATURAL MONSTERS

In 1881 Dr William Guy, former medical superintendent of Millbank Penitentiary, recommended lifelong detention for the estimated 131,000 recidivist thieves, prostitutes, vagrants, arsonists, vandals and street drinkers in England and Wales on the nod of two prison medical officers. His proposal received short shrift from other experts. It seemed so very illiberal. In England, we imprison an individual because they have committed a criminal act – not because of the 'type' of person s/he is, surely. However, around the same time, William Hardman, Justice of the Peace and chairman of the Surrey Quarter Sessions, expressed the wish that the authorities would devise an institution, 'half-workhouse, half-prison', where prisoners who had been observed as 'imbecilic' could be detained indefinitely after their prison sentence ended:

> You see, these people I speak of are a long way from the imbecile who goes into an asylum of that sort. They are just on the borderland, and I think they would require different treatment, a much more coercive treatment than the imbecile pure and simple.

Hardman said he had no concerns whatsoever about the liberty of the subject when the criminal classes were under discussion:

> There is a great deal of nonsense talked about the liberty of the subject in this country . . . I have no doubt that if it were brought before parliament

there would be a great outcry, because the liberty of the subject was about to be interfered with.

Hardman's was an interesting early opening salvo in the fight against liberty – and it was the very kind of statement that ought to have inflamed a body set up to contest such announcements: the Personal Rights Association. The PRA had formed in 1871 to publicise and protest against any abuse of individual rights brought to its attention; sadly, it proved to be something of a paper tiger, and failed to get a larger conversation under way about the threat of indeterminate sentencing or preventive detention *à la* Guy and Hardman. It would remain an insular coterie that preferred to fight small, one-off cases and lacked the energy and broader vision that would have made Hardman's statement the starting point of a debate about personal liberty. Nor did the PRA rally opposition to two significant changes in the relationship between the citizen and the law: the 1869 Habitual Criminals Act and the 1871 Prevention of Crimes Act mandated previously unthinkable levels of police surveillance of citizens and the latter created a 'sus' law. The Acts also led to the creation of the Habitual Criminals Register, which listed every person in the nation who had been convicted at least twice.

In 1898 the Inebriates Act enabled the indefinite detention of habitual drunks (persons arrested four times within one year for being drunk and disorderly). This gave local authorities the power to hold chronic drunks in a reformatory for up to three years; in the event, only London and Lancashire took advantage of these powers, and the experiment did not last long. Then, in 1908, came the big one: the Prevention of Crime Act of that year permitted the courts to add on an extra term of between five and ten years when a habitual offender came to the end of their most recent sentence. This was new. This was concerning. But there didn't appear to be any great agitation against these illiberal late-Victorian/Edwardian measures in comparison to other, earlier battles, notably the nationwide movement to repeal the Contagious Diseases Acts of the 1860s.

In an attempt to bring down the infection rates for venereal disease, the Contagious Diseases Act of 1864 gave police the power to detain, inspect and force into a specialist hospital any woman suspected of being a street prostitute and therefore a possible carrier of venereal disease. There was a particular anxiety about the number of soldiers and sailors rendered unfit for service after catching such an infection, and while initially only garrison towns were covered, two subsequent Acts extended the remit to other locations. The male client – who played an equal role in transmission – was not subjected to any scrutiny.

The Contagious Diseases Acts fuelled the burgeoning feminist movement, and after an epic campaign by women's rights activists, parliament repealed the legislation in 1886. Women of all classes, and many sympathetic men, had joined together to fight this clearly unjust set of measures – targeting poor women. But the rights of alcoholics and of petty offenders (and suspected petty offenders) did not spark any similar level of protest. For the most part, the press and the public at this time tolerated measures presented as benefiting the many at the expense of the antisocial or 'abnormal' few. The individualist libertarians, so keen to keep the state and the law from making incursions into citizens' lives, baulked when it came to the most undesirable elements of society.

***

The prison population of England and Wales had hit a peak in the 1860s, with 30,000 men, women and children incarcerated in either convict prisons or local prisons. After that, the increased use of fines, together with the expansion of other types of punitive institutions – especially for children – saw the number of prison inmates decline. Between 1885 and 1898, for example, the prison population of England and Wales fell by twelve per cent. Children and adolescents were now being sent to 'industrial schools' and 'reformatory schools' and, after 1908, into the new borstals. Juvenile courts were also created in 1908; and the use of probation had grown from the 1880s onwards. The broad trend in the second half of the nineteenth century was towards greater classification of types of offender, with particular urgency given to removing children and adolescents from the adult prison and judicial system.

Historian Stephen Watson has identified the key role played by the figure of the prison medical officer in classifying the mental state of inmates, from the 1860s onwards. One of the medical officer's duties was to spot the malingerers who hoped to evade justice by using the insanity defence, isolating them and observing their behaviour in order to distinguish them from the genuinely insane. The medical officer also had to assess the extent to which the prisoner was fit to receive a variety of physical and psychological punishments. Inmates who failed to respond to punishment were categorised as 'unfit for discipline'; these were the 'weak-minded criminals' with little ability to control their anti-social impulses. In this way, the problem of recidivism became linked with mental weakness, and the medical officer advised on the segregation of the physically and psychologically 'abnormal' from the rest of the prison population. In other words, the very fact of reoffending, of therefore being a 'habitual offender', was now conflated with 'weak-mindedness' – being unable to learn from the experience of punishment and being unaware of, or indifferent to, the expectations of behaviour from wider society. At the end of the century the label 'feeble-minded' would largely take over from 'weak-minded', and both terms were synonymous with 'unfit for discipline' within the penal system. (Neither 'weak-minded' nor 'feeble-

Newgate Prison, 1872: a man has just been 'flogged' with the cat o' nine tails whip. One of the prison medical officer's tasks was to decide which inmates were 'unfit for discipline' because they were 'weak-minded'.

minded' would ever firm up into a precise diagnosis – over the coming decades, these adjectives would be used to describe people with a broad range of psychological and social problems, and even those who used the terms liberally would admit their vagueness and imprecision.)

In addition to everyday observation of inmates, prison medical officers had access to meaningful paperwork that could trace the life, crimes and social background of those who were suspected of being weak-minded; as we shall see later, this detailed case history documentation was lacking in most other settings, severely hampering the quest to assess whether it was biological inheritance or the cultural environment a child was born into that caused their criminality. The bureaucracy that flourished within the nineteenth-century judicial and penal systems led to the compilation of relatively sophisticated case histories that both eugenicists and environmentalists would pore over.

Within the prison system, from 1876, the previous convictions of those sent to gaol for their second offence were noted down on a form devised for this very purpose; educational attainment, physical attributes and behavioural peculiarities were also recorded.

Unlike the moral insanity diagnosis of earlier in the century, the recidivist/weak-minded condition was not a disease that attacked a previously 'sane' person, but was instead believed to be a congenital defect – present from birth, permanent and incurable. Doctors deemed the 'morally insane' individual of the high Victorian age to be cunning and able to disguise the condition for long periods; the weak-minded of the later nineteenth century, by contrast, exhibited repetitive antisocial behaviour or 'social incompetence', which they had no guile to conceal.

There remained an unresolved conundrum, regarding the 'moral defective' category: many people who would find themselves labelled in this way would in fact have at least average IQ and educational attainment, while others would seem to be incapable of behaving well because they were too unintelligent to understand the world around them. When the term 'psychopath' was coined in 1885, it was noted that this type of personality showed intelligence, rather than otherwise. Forensic psychiatrist Charles Mercier, who would have considerable input into formulating the moral imbecile category, described psychopaths as 'clever fools'.

Reconciling these two phenomena would prove difficult in the century to come, as we shall see in a later chapter. Many doctors would only certify someone as a moral defective if they had a well-documented history of limited intelligence in childhood or adolescence. However, in the case of persistently aggressive or anti-social individuals with above-average intelligence, medical men would often refuse to certify under the Mental Deficiency Act. Medical officer Allan Warner wrote to the *British Medical Journal* in 1927, making this very point:

> Many feeble-minded persons, owing to their defect, commit anti-social acts and are quite properly certified as feeble-minded persons. But there is a class of person who is morally defective without any intellectual defect. Such a patient I do not consider certifiable as a feeble-minded person. It is true they require care, supervision and control like the feeble minded; but the feeble-minded person is, by definition, related to the imbecile and

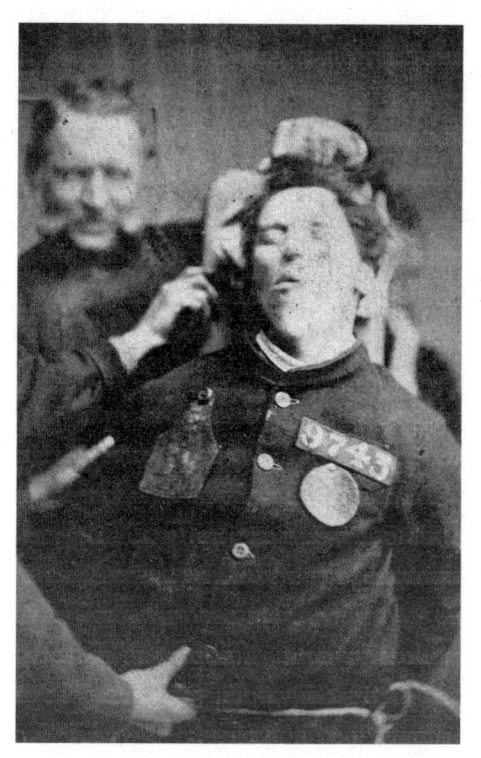

Wakefield Prison, 1869: prisoner 9743 struggles
against the punishment about to be inflicted.

idiot – it is only a question of degree. The moral defective who has no intellectual defect cannot be said to be related to the imbecile, and he should not be certified as, or treated with, the feeble-minded.

Nevertheless, other doctors would claim that persistent anti-social behaviour, particularly if it showed aggression, or its opposite, inadequacy, was evidence of 'incomplete emotional development', and therefore did represent a form of mental defect, regardless of how intelligent the individual proved to be.

Between 1879 and 1895, three commissions recommended that separate provision be made within the prison system for those found to be 'weak-minded'. These were the inmates who were not criminally insane (the criminally insane were destined, after 1863, for Broadmoor Hospital) but whose behaviour ranged from eccentric through to destructive and even suicidal. Estimates put the 'weak-minded/feeble-minded' at around three per cent of the prison population, with male prisoners twice as likely as female to be placed in that category. The 1898 Prison Act ensured that weak-minded prisoners were indeed segregated from others according to the judgment of the prison medical officer, and were concentrated at Parkhurst Prison.

***

Although prison medical officers noted down physical characteristics in their inmates' case histories, they tended not to adhere to the idea that physiognomy, cranial dimension or any other anatomical feature was indicative of criminality, despite the fashion for such a belief among certain sections of the intelligentsia. Cesare Lombroso's 1876 work *L'Uomo Delinquente* (*Criminal Man*) isolated the 'born criminal' (or 'congenital criminal') as 'an organic anomaly' who was physiologically different to non-criminals. These differences, Lombroso theorised, indicated born criminals' links to more primitive forms of hominid: they were less securely evolved than their fellow, non-criminal, human.

Though *L'Uomo Delinquente* would not be published in English until 1911, Lombroso's views were popularised in this country long before then, not least by sexologist Havelock Ellis. In his 1890 work *The Criminal*, Ellis stated his belief that criminality had two aspects: innate disposition, and environmental contagion. Ellis admitted that social science was not sufficiently advanced to be able to untangle which of these two had the greatest influence. For Ellis, the insane, the epileptic, the cerebro-spinally diseased, the overworked, the neurotic and older fathers were disproportionately producing criminally inclined offspring – though he also thought environmental factors played a role.

Ellis believed that by imprisoning the 'occasional criminal' (who was not an abnormal type) the penal system simply 'manufactured' the habitual criminal by a slow and subtle process: the prison system was 'like a sewer', he wrote, flushing hardened, more sophisticated broken people out into society. The instinctive criminal, however, was a very different phenomenon – a 'natural monster' (who seems close to our modern understanding of the concept of the psychopath):

> It must be remembered that the lines which separate these from each other, and both from the instinctive criminal, are often faint or imperceptible... In the habitual criminal, who is usually unintelligent, the conservative forces of habit predominate; the professional criminal, who is usually intelligent, is guided by rational motives and voluntarily takes the chances of his mode of life; while in the instinctive criminal, the impulses usually appear so strong, and the moral element so conspicuously absent, that we feel we are in the presence of a natural monster.

Experiments in anthropometric measurements to link crime to heredity were, as already stated, only patchily accepted in Britain, and for every alienist excited by the Lombroso approach there were many who could not be convinced; and so anthropometry struggled to be seen as anything other than a pseudoscience.

***

By the turn of the twentieth century, not only was the prison population declining, but reported crime was also falling in almost every category, according to the statistics collected each year by the Prison Commission. The number of people who could be described as hailing from 'the criminal classes' (as the Prison Commission termed them) had shrunk by one-quarter since the early 1870s. One of the few offences that had seen a significant and steep increase was a new crime – the contravention, by parents, of the Elementary Education Act of 1870. And thus another arena in which alarm began to set in, with regard to the mental capabilities of the British, was the schoolroom.

3

'EDUCATION ON AN EMPTY STOMACH'

The Elementary Education Act of 1870 created a national system of state-funded schools, partly in recognition that Britain would need a fully numerate and literate population if it was to keep pace with the explosion in clerical and technical jobs, which required vast reserves of white-collar workers. In Germany, often the comparator nation in these years, an educated working class was efficiently servicing the modern world of work. If Britain did not keep up, the prosperity of the nation was likely to suffer. 'National efficiency' and 'social efficiency' were the increasing obsessions among policymakers, and would play a major role in the arguments about the allegedly mentally defective.

Most British children had attended a school of some kind before the 1870 Act – usually one of the Church of England's network of 'National Schools' or the religious Non-Conformists' 'British Schools'. But the School Boards created by the 1870 Act brought greater national standardisation of curricula. Some parts of the country made elementary schooling compulsory; and a follow-up Act of 1880 removed any leeway, compelling all parents of children between the ages of five and ten to ensure that their children attended school full-time; and at least as 'half-timers' between the ages of ten and thirteen. (Half-time schooling would be abolished in 1900.)

As a result, for the first time, vast numbers of British children were collected together, in the huge new Board Schools that were constructed, often in the heart of deeply deprived districts; and, as with the large asylum and prison populations who were similarly concentrated on single sites, the surveying, classification and analysis of the 'inmates' became possible as never before. Meaningful data could now be collected on the physical, intellectual and 'moral' (i.e. behavioural) state of the nation's children. While standardised tests for intelligence, memory, vocabulary, deduction and spatial perception would not be developed for British schoolchildren until the early twentieth century, from 1870 many teachers and School Board administrative staff sounded the alarm about the scale of pupils' 'backwardness', as well as neglect, exhaustion, starvation and disease. In 1880, just over half of schoolchildren failed to achieve the standards that had been set by the School Boards.

In Bradford, future Independent Labour Party politician Fred Jowett would persuade Bradford City Council in 1904 to provide free school meals for the most famished pupils. Bradford was the first English local authority to do so; in 1893 Bradford had been first to appoint a school medical officer, and in 1908 would open the country's first school clinic. 'Education on an empty stomach is a waste of money', Jowett declared.

In London, where the School Board had adopted compulsory attendance as early as 1871, vivid eyewitness accounts, such as the following, suggested the impact of a slum environment on the mental capacities of the very young:

> We see numbers of half-imbecile children throughout the school; big boys in low standards who cannot learn, try as they may; children of drinking parents chiefly. Sometimes a boy is running wild for weeks together . . . The girls are anxious-eyed, with faces old beyond their years . . . Abstract thought dazes them, but they are shrewd enough when the subjects are things they know about . . . Cookery classes were introduced after pressure . . . [and] the latent womanliness is developed.

> Puny, pale-faced, scantily clad and badly shod, these small and feeble folk may be found sitting limp and chill on the school benches in all the poorer parts of London . . . The practised eye can readily distinguish children of this class by their shrinking or furtive look, their unwholesomeness of aspect, their sickly squalor, or it may be by their indescribable pathos, the little shoulders bowed so helplessly beneath the burden of the parents' vice.

HOLLAND ST. BOARD SCHOOL, BLACKFRIARS.
Lowest type; present time. Compare with Lant St., 1875, and note improvement.

LANT ST. BOARD SCHOOL (SOUTHWARK), 1875.
Lowest type.

LANT ST. BOARD SCHOOL (SOUTHWARK), 1878.
Slight improvement.

London schoolchildren were surveyed to explore whether ongoing physical and mental 'degeneration' was taking place in impoverished districts.

Just how many children were 'dull', 'slow', 'feeble-minded' or, in fact, so 'mentally deficient' as to be in need of special provision outside of the ordinary elementary school was difficult to assess accurately without testing that had agreed, standardised criteria. For what it's worth, a 1905 governmental inquiry estimated that one per cent of children at elementary schools in England and Wales could be described as 'mentally deficient', while the London School Board gave the figure of 0.5–0.6 per cent of the capital's circa 800,000 school-age children as too 'mentally deficient' to be catered for by ordinary elementary education (i.e. around 4,000–5,000 pupils).

Separate provision for four categories of schoolchildren was introduced in the 1893 Elementary Education (Blind and Deaf Children) Act, and the 1899 Elementary Education (Defective and Epileptic Children) Act; these Acts permitted local education authorities to segregate such children from 'ordinary' pupils, if they saw fit, and to place them either in separate classes or in 'special schools'. If the Acts were locally adopted, parents of defective or epileptic children were obliged (under threat of a £5 fine) to bring forward their child to be examined 'by a duly qualified practitioner' if it was suspected that s/he was 'incapable of receiving proper benefit from instruction in the ordinary public elementary schools, but not incapable of receiving benefit from instruction'. This latter phrase meant that the child was not so 'defective' as to be classed as an 'idiot' or an 'imbecile' – i.e. almost entirely ineducable and in need of a wholly different type of segregation and care.

This assessment or 'ascertainment' of schoolchildren was supposed to distinguish a child who was not merely a little slow in learning, nor so disabled that they could learn nothing, but who fell instead into the new category somewhere between the two – 'mentally defective'. The London School Board named these the 'borderland cases'; and this border territory was hugely contentious – a spectrum of educability and improvability about which few could agree during the first half of the coming century. (Chapter 11 will return to the topics of testing, IQ and cultural factors affecting learning.)

When the London County Council (LCC) took over the duties of educating young Londoners from the London School Board in 1904, the LCC became one of the most proactive and energetic bodies in the country in searching out and ascertaining children as 'mentally defective'.

4

'HUMAN DREGS AT THE BOTTOM OF OUR NATIONAL VATS': THE COLLAPSE IN THE MIDDLE-CLASS BIRTH RATE

On 16 October 1903, just two months before Eichholz made this radical statement, eugenicist Professor Karl Pearson had given the Huxley Lecture, entitled 'The Laws of Inheritance in Man', in which he dramatically stated the hereditarian view that the 'wrong' type of Briton was significantly outbreeding their social superiors:

> We are ceasing as a nation to breed intelligence as we did fifty to one hundred years ago. The mentally better stock in the nation is not reproducing itself at the same rate as it did of old; the less able and the less energetic are more fertile than the better stocks. No scheme of wider or more thorough education will bring up, in the scale of intelligence, hereditary weakness to the level of hereditary strength. The only remedy, if one be possible at all, is to alter the relative fertility of the good and the bad stocks in the community.

A fall in the birth rate had taken place right across Europe from around 1870, and in Britain, between 1871 and 1901, it fell by twenty per cent; the middle and upper classes had seen the sharpest drop. The big decline among the wealthier Britons was caused by the more frequent use of both 'natural' and mechanical contraception; a rising age of marriage for men, who increasingly sought to establish an income and a position in life before proposing; and broadening educational and professional or voluntary work opportunities for women, meaning that marriage and children no longer had to be their most likely fate. The educated, urban 'New Woman' at the end of the nineteenth century was turning her back on wifedom and motherhood. The revelations in the newspaper press of Divorce Court hearings – domestic violence, infection with venereal disease by a promiscuous husband, the stultifying life of the homebound wife – assisted many women to make up their mind to remain single and childless. The very women eugenicists would have wanted to pass on their qualities were choosing in greater numbers not to do so – it was a dilemma for (male) eugenicists.

There wasn't anything new in elite alarm at the seeming super-fecundity of the poor. As historian David Eversley has stated, 'The demographic decline of ruling castes is part of the general tradition of Western social thinking.' What gave it fresh impetus in the final decades of the nineteenth century and the start of the twentieth was the emphasis on the biology of the social classes, together with a notion shared by many that the state's intervention was rescuing the pauper class from deserved extinction. Such 'kindnesses' as the Poor Law, with its workhouses, infirmaries and 'Outdoor Relief'/'Out Relief' (that is, cash, bread, coal, etc., delivered to the pauper's home); public health measures; sanitation law; institutions such as lunatic asylums, schools, lock hospitals for venereal disease, inebriate homes, 'idiot' asylums and so on, had all artificially kept alive, and fertile, those who would have been eliminated naturally as a result of being 'unfit'. In effect, the public purse was now supporting the poor and the sick to produce defective stock. Later, H.G. Wells would put it like this, addressing his imagined members of the underclass who opposed the 'modernity' of eugenics:

> We cannot go on giving you health, freedom, enlargement, limitless wealth, if all our gifts to you are to be swamped by an indiscriminate torrent of progeny. We want fewer and better children who can be reared up to their full possibilities in unencumbered homes, and we cannot make the social life and world peace that we are determined to make, with the ill-bred, ill-trained swarms of inferior citizens that you inflict upon us.

Such attitudes faced angry opposition. The nation's medical officers of health were among the most vocal anti-hereditarians and were in the best position to appreciate the real impact of public health interventions. Arthur Newsholme, chief medical officer at Whitehall's Local Government Board, for one, saw co-operation, not competitiveness, as a higher human evolutionary trait. He vehemently disagreed with withholding help to the weak and the chronically poor, mocking Darwinists as no doubt opposing sunshine, since it kept so many of the 'unfit' alive. Other interpretations of Darwin were possible, though: geologist, anarchist and evolutionary thinker Peter Kropotkin travelled to Siberia with zoologists to observe animal behaviour there and later, in 1902, wrote, 'We vainly looked for the keen competition between animals of the same species which the reading of Darwin's work had prepared us to expect.' Instead, Kropotkin saw co-operation and altruism. This formed the foundation of his own political creed of Mutualism, in which humans co-operated to achieve economic output and shared its rewards. Kropotkin liked to quote this passage from Darwin's *Descent of Man* (1871), to thumb his nose at those who had interpreted Darwin's work as supporting the selfish, hard-hearted, Survival of the Fittest ethos: 'Those communities which included the greatest number of the most sympathetic members would flourish best, and rear the greatest number of offspring . . . We are impelled to relieve the sufferings of another, in order that our painful feelings may be at the same time relieved.'

Nevertheless, the term 'race suicide' entered the discourse, escalating the sense of panic. To explore the nature of the demographic changes, the Census of 1911 included additional questions on the rate and number of births to married couples. The results of this so-called 'fertility census' would not be fully analysed and published by the Registrar General until 1917 and 1923, as the two-part *The Fertility of Marriage Report*. The findings confirmed that the upper-middle and professional classes had taken the greatest steps to control their fertility; next came the skilled manual workers; while the unskilled were least likely to try to limit family size.

In the meantime, eugenicists had continued to undertake their own surveys, and in 1909 Dr Ettie Sayer and Dr Alfred Tredgold submitted their separate figures to the *Eugenics Review* magazine on feeble-minded female fertility. Sayer claimed that the average married couple in London produced five offspring while the feeble-minded London woman produced 7.6 offspring. Tredgold, for his part, put the respective figures at four and 8.4, and wrote that 'many of these women, sometimes even mere girls, are possessed of such erotic tendencies that nothing short of lock and key will keep them off the streets'.

But erotic tendencies are how offspring are created – Nature's way of getting life-forms to beget more life-forms: why could Tredgold et al. not see that? Why could they also not see that a child reaching adolescence in the slums – surviving the extraordinarily tough physical conditions of chronic poverty, and blossoming into adulthood – served to prove that *they* were the 'fittest'? Hadn't Darwin himself written that the best-adapted creature is the one that gets to propagate its own kind?

***

An estimated 0.5–1 per cent of schoolchildren and three per cent of the prison population were 'mentally deficient'; and the Census of 1901 put the total weak-minded population of England and Wales at 0.14 per cent. On the face of it, this didn't quite add up to the demographic emergency that eugenicists were insisting upon. But the phenomenon that perturbed them most – and which they would successfully conjure up to terrify the policymakers – was the feeble-minded woman pumping out litters of equally feeble-minded offspring. So powerful was this bogeywoman of the mental deficiency debate that we will give her a chapter of her own.

5

'ONE LAW FOR MEN, AND ANOTHER LAW FOR WOMEN'

'It is an exception for any of them not to have been sexually tampered with', declared Dr Thomas Clouston, physician superintendent of the Royal Edinburgh Asylum, of the many 'feeble-minded' female inmates whose life stories he had been able to trace. He continued:

> Many of them have had illegitimate children, and this often at a very early age. One had seven such children. I look on this source of immorality as an extremely grave one in our social life. In a way, it is more disgusting and degrading than prostitution or sexual lapses through passion. When illegitimate children are borne by such women, the chances are enormously in favour of their turning out to be either imbeciles, or degenerates, or criminals.

This borderline category of female haunted the case-books of Poor Law Guardians, workhouse staff, medical superintendents, 'rescue' societies, reformatory schools, industrial schools, certified schools – the panoply of institutions of rescue and reform, punishment and protection. Everyone had plenty of stories about their fecundity, their vulnerability, their promiscuity, their stupidity, and it's worth re-telling a selection of these for what they reveal about girls' and women's experiences and the anxieties they evoked in people with oversight of their lives.

Anecdotes concerning illegitimate babies were commonplace, recounted in order to underline the out-of-control breeding by females who were simply not bright enough to stop it happening to them. Baldwyn Fleming, a Local Government Board inspector with many years' experience of the workings of the Poor Law throughout England and Wales, despaired of destitute girls and women of weak intellect giving birth repeatedly to children who had comparatively little chance of survival to adulthood.* In 'C Workhouse', Fleming said, anonymising his material, he knew of twenty-five-year-old D— F—, who was 'physically and mentally defective', though the local medical officer would not certify her under either the Lunacy Act or the Idiots Act – she clearly was not so severely lacking in mental capacity as to meet those thresholds. According to the workhouse matron, 'DF' had 'no idea what to do with her baby . . . a poor undersized little object'. (This is *Oliver Twist* language, seventy years after Dickens's tale.)

The matron's efforts saved the baby, and the matron had hoped to be able to keep the child and 'DF' in the workhouse long enough for them both to be able to build up their strength. But as 'DF' was over the age of sixteen, the institution could not detain her against her will – she only had to give twenty-four hours' notice before walking out. 'DF' expressed the wish to leave with her baby, and 'what would be the almost inevitable result?' asked Baldwyn Fleming. 'That in a few more months, the child would be dead, and the mother would again be pregnant.' (Except in rare circumstances, when the Poor Law Guardians could formally 'adopt' a friendless adolescent until age eighteen, boys and girls were free to leave the workhouse on reaching sixteen. And by sixteen, to the authorities' chagrin, sexual maturity and activity would have probably commenced.)

In another workhouse was F— A—, aged thirty-three, 'weak-minded', who had had a child at seventeen, and had since had four more – three of whom had died as babies. 'She is quite unable to protect herself from the danger she incurs when free from restraint', said Fleming.

Fellow Local Government Board inspector Philip Bagenal (whose remit had included the east of England and Yorkshire between 1896 and 1905) revealed similar instances, in which he and the Poor Law authorities

---

* In 1900, an illegitimate baby was twice as likely to die under the age of one as a child born in wedlock, with the deaths attributable to poverty, neglect and the practice of 'baby-farming' – the boarding out of infants to an unregistered childminder.

felt powerless to deter (or protect) pauper girls and women from repeated pregnancies with illegitimate babies they believed were unlikely to thrive. When pushed for some statistics, Bagenal revealed that of a population of 112,365 in Wakefield, some fourteen pauper females on Poor Law Outdoor Relief were feeble-minded; in Sheffield, of a population of 246,000, he put the figure at forty-one. These are low proportions, but they caused disproportionate unease: 'Of course, it seems obvious that any person of this class who is in receipt of Outdoor Relief, a female especially, is more or less a source of danger to the community', Bagenal stated. She was seen a danger while also being in danger herself.

Bagenal knew of a forty-five-year-old unmarried workhouse woman with three children:

> She is on the borderline of imbecility, but the medical officer will not certify her as insane. Two of the children are said to be the result of an incestuous connection. She makes no application to go out, but there is no means of legally detaining her . . . I find in the infirmary or hospital wards [of the workhouse] in 1904 there had been three cases of illegitimate births, all the offspring of feeble-minded women. One was the result of an incestuous connection. The superintendent nurse told me that the mothers of all these cases were distinctly feeble minded, and her experience is that illegitimacy is very frequently the result of this.

Robert Parr, director of the National Society for the Prevention of Cruelty to Children (NSPCC), also expressed alarm, in 1905, at girls being impregnated by family members. He cited a case he had dealt with in Manchester of a young woman who lived with her two brothers, by whom she had four illegitimate children. 'All [siblings] are of weak intellect. The first child lived three days and died in a fit. Second child stillborn. Third lived a year and four months and died of natural causes. Fourth lived five weeks and died in workhouse.' The NSPCC officer had taken the young woman and her eldest child to the workhouse as a place of safety but she discharged herself, and returned to her brothers' home. Twenty-four hours after giving birth to her fourth child, the local doctor discovered her in bed with one of the brothers. (Pressure by the NSPCC, together with the National Vigilance Association, the Ladies' Association for the Care of Friendless Girls, and other women's rights and 'social purity' campaigners, would see the Punishment of Incest Act passed in 1908, making incest a specific criminal offence for the first time, though prosecutions could take place, before 1908, under existing rape and sexual assault laws.)

Many experienced staff did not hesitate to point a finger at the depravity of 'normal' men who had intercourse with a female who they knew, or suspected, was unable to give consent due to intellectual weakness – even supposing that her consent had been sought at all, and that this was not simply rape. Philip Bagenal believed 'feeble-minded women are particularly open to the seductions of men. They seem to be deficient in will power and the power of resistance to attacks upon their virtue.' However, Bagenal could not stop himself splattering some of the blame back on to the woman, where 'in some, the moral sense is altogether absent.'

This was not an uncommon view, and many seemed unable to confront the issue of rape and indecent assault of 'feeble-minded' women and girls without equivocating, and trying to confect a case against the victim – be it naivety, stupidity or 'immoral habits' (a phrase used by the NSPCC and the LCC). They assumed she had contributed to her own assault and impregnation. So did Dr John Sutherland, Deputy Commissioner in Lunacy in Scotland, who also believed that like bred like:

> Again and again I have seen in the same district, and even under the same roof, three generations of 'soft' or 'feeble-minded' women with illegitimate broods, some of whom are imbecile or 'soft' or feeble minded . . . Among these broods of illegitimate children there is, in proportion, a vastly larger number of imbecile and feeble-minded children met with than among the offspring of normal and healthy wedlock . . . One faux pas is intelligible, and if all the circumstances were known, perhaps excusable; but repeated breaches of the moral law, and all the hardships it entails on the woman, not to speak of society, is suggestive of feeble-mindedness of the female parent. The feeble-minded male is not an offender in this respect; it is the female taken advantage of by unscrupulous men, and the female who is actively erotic.

Maria Poole did not agree though: Poole had been for twenty years the secretary of the Metropolitan Association for Befriending Young Servants – a 'rescue' society that took in girls from Poor Law institutions, industrial schools, the NSPCC and other voluntary bodies, and trained them to go into domestic service. Because of the eternal 'servant problem', such labour was always required, even in times of high unemployment. The 'MABYS', as they were known, dealt with around 600 girls at a time, across Britain, and in Poole's experience, feeble-minded servant girls very often had 'lost their character' through 'want of self-protection, I think, very much more than wilful sin.' They had experienced 'contamination by men . . . Their lack of mental power prevented them from being self-protecting.'

Poole shared a common frustration that the medical profession would not certify feeble-minded girls and women as idiots or imbeciles; this would, Poole argued, have given them the protection of the lunatic asylum or idiots' home. (Despite the passing of the 1886 Idiots Act some people with learning disabilities were held in lunatic asylums, often because of a shortage of idiot home places.) Feeble-mindedness, though, as we have already seen, was not a diagnostic category, and medical men proved reluctant to certify; they would have had to make a case that a girl or woman was less mentally capable than she actually was.

She expressed despair about the existing legal safeguards. The Criminal Law Amendment Act of 1885 had made it an offence for anyone to 'unlawfully and carnally know any female idiot or imbecile woman or girl, under circumstances which do not amount to rape, but which prove that the offender knew at the time of the commission of the offence that the woman or girl was an idiot or imbecile'. (The punishment for this was up to two years in gaol, with or without hard labour.) But proving before a jury that a male had been aware of a female's mental capabilities was extremely difficult; and when such cases came to court, her evidence was all too easy to undermine precisely because her intellect was not strong. A judge might even decide that no trial could proceed, as the alleged victim was not a competent witness. 'I do not see myself what you could do unless you gave him heavier punishment', said Maria Poole, 'but he is very often not found [guilty].'

The NSPCC was determined that the Criminal Law Amendment Act should be made a lot more effective with regard to 'feeble-minded' females. The Society's Robert Parr claimed:

> We are meeting almost every other day with cases in which girls are taken advantage of by able-bodied men, and it is absolutely impossible to enforce... the Criminal Law Amendment Act because the girl has not wit enough to know quite what the man did, or who the man was. It is impossible to enforce the Act because you cannot clearly prove that the man knew that the girl was an imbecile. I think the responsibility should be put on the man to find out before.

(As we will see later, that last suggestion would be enacted, in 1913.)

In November 1906, thirty-seven-year-old Frederick Briden was found guilty at the Old Bailey of having non-lawful intercourse with Jane De Vries, nineteen, whom the court accepted as an 'imbecile within the meaning of the Criminal Law Amendment Act'. De Vries was in fact a 'deaf-mute' and her status as an 'imbecile' was pored over by prosecution and defence. Eventually they decided that legal case law had established that 'deaf-mutes' were included in the legal definition of imbecility, which they declared was 'a collection of indefined qualities, childishness being amongst them'.

The judge told the jury that they should consider whether De Vries had such weakness of mind that 'under the will of the man, there was no chance for the woman – the woman must be incapable of resisting persuasion, of exercising an act of her own will, or of giving or withholding her consent'. They should ask themselves two questions: was the woman an imbecile? And did the prisoner know that she was? The jury answered yes in both cases, finding Briden guilty, and the judge sentenced him to four months in prison without hard labour – a sentence length that will come as no surprise to present-day readers.

Other Old Bailey sentences handed down between 1885 and 1910 for men found guilty of carnal knowledge of an 'imbecile' female range from Briden's four months without hard labour to twelve months with hard labour – which was one year less than the maximum sentence permitted in law. The custodial sentences for the victims, and potential victims, of such attacks would be a great deal longer, once the Mental Deficiency Act was passed.

***

Ellen Pinsent was worth listening to: her long experience of, and noteworthy intervention into, the lives of the learning-disabled gave her huge authority. In 1896 Pinsent had co-founded the National Association for the Care and Protection of the Feeble-Minded, which set up a number of small charitable refuges for girls, who, she acknowledged, were not sufficiently well cared for or protected by the workhouse, reformatory, industrial school and so on. Pinsent would be one of the middle- and upper-middle-class women who at the *fin de siècle* moved successfully from the philanthropic sphere into local government, policymaking and eventually the Civil Service. Pinsent became chair of the Special Schools Subcommittee of the Birmingham Education Committee and joined the Eugenics Education Society shortly after its foundation in 1907.

In a 1903 article for *The Lancet*, Pinsent alluded to the fact that specialised permanent detention would be kinder on the public purse:

> Some of them are girls who should never be allowed out of the house by themselves, instead of which I know that they are prone to wander alone for hours in the streets. A poor mother with other children cannot give her whole time and strength to the weak-minded member of the family. The child must take her chance during the busy hours of the day. These girls are often of a clinging and affectionate disposition and will follow any man who chooses to speak kindly to them. We who pay the rates and taxes will have to support these girls in the end. Would it not be wiser to do so at once, instead of waiting until they have produced others for us and our children to support?

Ellen Pinsent's influence on Birmingham's educational and social work provision was mirrored in Manchester by Mary Dendy – her co-founder of the National Association. Dendy set up a large charitable institution for feeble-minded children and adults at Sandlebridge in Cheshire, having founded the Lancashire and Cheshire Society for the Permanent Care of the Feeble-Minded. At Sandlebridge, strict sex segregation was enforced, and permanent, lifelong care aimed at. Ellen Pinsent was aghast at the release of inmates from rival organisations after a few years of training; she believed that it was wrong to pretend that any feeble-minded people were 'normal'. Dendy worried most of all about the 'borderline' people

– the very 'high-grades' who shaded off into 'normal'. In 1910 she wrote, 'It is the mild cases, which are capable of being well veneered, so as to look, for a time, at any rate, almost normal, against which there is most need to protect society.' This is fascinating, as it echoes earlier concerns by Victorian psychiatrists about the highly eccentric who may be walking among 'normal' people, successfully suppressing their 'hidden madness'. It was an unsettling notion that fed Sensation Fiction and the emergent horror and mystery fiction genres, from the 1830s onwards.

Ellen Pinsent, left, and Mary Dendy were pioneers of the new custodial approach in learning disability. Both women believed that 'mental deficiency' was biologically transmitted and that defectives should not be permitted to have children.

Pinsent and Dendy strongly advocated for permanent detention of the feeble-minded as a new way of tackling a number of social issues all at the same time: to protect such individuals from sexual predators and from other types of predation too, for example, financial fraud and exploitation of their labour; to relieve the overcrowding in such state institutions as asylums, prisons, industrial schools, reformatories and workhouses; to bring down crime rates; and – as both women were keen hereditarians – to prevent the breeding of future 'mental defectives'.

The National Association for the Care and Protection of the Feeble-Minded had a membership that was overwhelmingly female. From 1907, nearly half of the early membership of the Eugenics Education Society, too, were female. Why did women take the lead, from the 1890s, in advocating the lifetime incarceration of their sisters? Why did they interest themselves to such a noticeable degree in the burgeoning eugenics movement? Well, in an era in which it remained controversial for a woman to take a role in public life, these particular socio-political topics were comparatively safe for women – for 'ladies' – to campaign upon, and to write and to speak about on a public platform. She remained pretty much above reproach when concerning herself with the lives of children, families and girls. When a female operated in such stereotypically feminine areas, traditionalists could find less secure cause to protest that a woman's sphere should be strictly the home or a limited range of philanthropic activities.

When eugenics came to prominence, motherhood became a hugely important topic, as historian Lesley Hall has pointed out. If the British 'race' (or indeed the human race) was to be revitalised through selective breeding, mothers unavoidably constituted half of this revitalisation process. Late-nineteenth-century feminists could see the link between eugenics and a greater status accorded to mothers and motherhood; this held true even among those who also believed that environment played as important a role as heredity in the 'quality' of the child. A significant number of male eugenicists favoured limiting women's lives to being good wives and breeders of superior human material. This created an obvious clash with feminism; but many women's rights campaigners comforted themselves that this clash was something that could be dealt with further down the line – once they had won the battle for suffrage and women's full access to higher education and the professions.

There was crossover, too, with the 'social purity movement', which sought to counteract (often in the form of legislation) male sexual behaviour that impacted adversely on the lives of girls and women. As noted in chapter 2, the sexual double standard came under increasing attack from the late 1860s, and male sexual licence was called out in agitations concerning prostitution, venereal disease, the age of consent and the prosecution of, and sentence lengths for, a range of sex crimes. Lucy Bland has written that 'genetic purity and moral purity converged in their construction of a new moral standard, and motherhood gained a new dignity as a "duty for the race"'. Eugenics was a realm in which women were central in bringing about 'the promise of a new reality'. Mothers were the 'race regenerators'.

Female eugenics activists were uniquely placed to get involved, with many of them having a long curriculum vitae in charity and social work that included compiling detailed case histories and organising philanthropic interventions in the lives of poor women and girls. The twentieth century saw many of these confident, articulate and dynamic women shift their field of operations from voluntary associations (with names including such terms as 'friends', 'army', 'league', 'ladies', 'vigilance', 'purity', 'protection', 'prevention') to local and national governmental bodies – their years of experience making them daunting individuals with whom to debate or to oppose. In Edwardian committees and sub-committees, Royal Commissions, departmental inquiries and so on, such women were accorded an authority and an agency rarely to be found in even the final decade of the nineteenth century.

One of the first women to qualify as a physician in Britain was Dr Ettie Sayer, whom we met earlier – a supporter of votes for women, a 'moral purity' campaigner (intent on curbing the spread of venereal disease to wives and children) and consulting physician to the LCC's Education Department. Dr Sayer wrote repeatedly to the Home Office urging the creation of a Royal Commission on prostitution; she believed that an open and in-depth governmental inquiry into the trade in flesh would shift the stigma from the supplier to the consumer and would be a first step in bringing down rates of venereal disease. She claimed that one in five Britons would suffer venereal disease at some point in their lifetime (a figure that is impossible to substantiate), and wrote in 1912 to Home Secretary Reginald McKenna that 'it is a byword among gynaecologists that "Gynaecology is Gonorrhoea"', referring to the terrible toll on wives' health when promiscuous husbands communicated their disease to them and their unborn children. Sayer was also a hard-line eugenicist and believed in the lifelong incarceration of both sexes of feeble-minded, as well as those she deemed 'morally defective', in order that they did not breed any more of their kind.

The journalist from *Truth* magazine who attended Dr Sayer's lecture and slideshow on 'morally defective children' at the inaugural

International Moral Education Congress in London in 1908 wrote that he had

> never seen anything more interesting . . . [She] showed us by the lantern, photographs of a normal, healthy brain and of a defective one . . . The first is crammed with memory cells, from which fibres run straight up to the surface of the brain and end in fine arborisations with fibres from neighbouring cells. The other has few cells, with small, shrivelled fibres, and not nearly long enough to get into touch with their neighbours. The unfortunate owner of this brain would be subject to violent passions, but devoid of reason, judgment and self-control. It ought to make us very charitable in our judgments, even of the worst criminals. *Tout comprendre c'est tout pardonner*. How can unfortunate men, women and children help themselves if they are born into the world with defective brains?

The *Daily Mail* also caught Sayer's lecture tour and gave her a headline: 'The Morally Defective Child – A Woman Doctor's Remarkable Lecture'. Sayer ended her presentation with the words:

> As to real moral degenerates . . . if diagnosed as so actively antisocial and morally indirigible as to be unfit ever to live among a pure, honest, unselfish and public-spirited people, they should be classified and shipped off to various uninhabited isles . . . and after they die, their brains should be placed at the disposal of the medical profession.

Sayer in one woman encapsulates the range of beliefs prevalent at this time: her professional life was dedicated to the protection of women and girls, and she broke ranks to promote a commonsensical confrontation of taboos, such as venereal disease, prostitution and child abuse. But then we see, with this clarion call regarding the 'morally indirigible', a harder-line and intolerant approach: such people are to be exiled and then anatomised.

***

To try to place Dr Sayer in context, educated women's support for eugenics cut across late-Victorian and Edwardian battle lines: feminist and non-feminist; conservative and liberal; conservative and socialist; religious and agnostic. Socialist women may have been expected to have sided with their outcast sisters by advocating state intervention into terrible living conditions and exploitative labour practices as the correct response to alleged race 'degeneration' – and most of them did indeed do so. But the majority additionally lent their support to state policies to discourage, or to prevent, the multiplication of the 'unfit'. For many socialists, eugenics additionally offered a critique of a dissipated, decadent aristocratic class – whose 'stock' ought to be replaced by sturdy, hard-working, rural types, untainted by over-refinement and excess. In the modern world, emancipated woman would be able to select her own healthy, intellectually and morally sound husband and father of her children; this method of sexual selection would rightly replace the upper-class marriage market that her mother and grandmother had had to endure. In this way, a meritocracy could emerge, where the best stock of the middle class and the 'respectable' highly skilled or white-collar section of the working class would replace the exhausted ancient lineages that had been busy interbreeding themselves to extinction, House of Usher-style.

There is no great cache of anti-eugenics feminist material from this era, though some oppositional writing can be found scattered across suffrage journals. For instance, this short editorial, entitled 'Eugenics and Women', appeared in *The Vote* magazine of 3 August 1912, in light of a forthcoming eugenics conference and imminent parliamentary activity concerning feeble-minded women in their fertile years of life:

> Women concerned with the freedom of their sex are warned to keep a watchful eye on the proceedings of the Eugenics Conference. In conjunction with the principle newly introduced by the government adoption of the Mental Deficiency Bill – a Bill as to which a warning note has already been sounded, on the ground that it will be enforced solely against women – this conference appears in the light of the greatest danger to women's liberties which has yet threatened them. One learned gentleman has not hesitated to say that the sole remedy for racial degeneracy is to be found in the control of feeble-minded women, entirely ignoring the heavy responsibilities of alcoholic, criminal and viciously inclined men, who, apparently, are to continue to go uncontrolled. And the dominant note of the

lengthy discussions has been the constant effort to fix the responsibility for all human ills on to the shoulders of women.

The Vote saw the potential danger of 'women being shut up in asylums at the fancy of a doctor, supported by a guardian. Not only can an unmarried mother come easily under its [the Bill's] provisions, while the father goes scot-free, but also those to whom "punishment is no deterrent". Suffragists, please note!'

Suffragists also noted the male-ness of the staff and administrators of the proposed mental deficiency legislation – the medical experts and the Poor Law 'relieving officers', 'with their ugly record of harsh and dictatorial dealings; officialdom unchecked and "science" in all the arrogance of "a little learning" and brief authority do not inspire women with unlimited confidence.'

Women were very slowly finding their way into such roles, including School Board positions, and as aldermen on the more enlightened local councils. But 'officialdom' was nevertheless overwhelmingly male – for the moment, at least. Women, unable to vote, had not been considered by the law-makers, the journal added.

Novelist and essayist Mona Caird was pretty much the only high-profile anti-eugenics feminist. She was to some extent ostracised for her outspoken attacks on female 'social purity' campaigners who went along with ideas about 'race improvement'. As literary historian Angelique Richardson has pointed out,

> Caird emphasised that the environment was a genuinely interactive force, 'the chemical union of native bias with daily circumstance which has for product a human character' . . . If the environment had a significant part to play, then the division of the poor into the deserving and undeserving was facile and redundant, and the idea of eugenics fundamentally flawed . . . [Caird] argued that social progress is dependent upon chance variation, not artificial selection.

From an unlikely source came objections to the sexual double standard of the proposed mental deficiency legislation. Sir Frederick Banbury, Conservative MP for the City of London, was no feminist; a strident opponent of votes for women, Banbury believed that the ladies had all the rights and protections they could want, thanks to the chivalry and paternalism then prevailing in England.* But the law had to be even-handed, he said, during the 1912 and 1913 parliamentary debates:

> Why not include men? Why should only women be brought under this Section [of the Bill]? I am serious upon this point, because I have always said that we are perfectly capable of looking after the interests of women but if we are to have such legislation as this, one law for men and another law for women who are incapable of due self-control or self-protection with respect to sexual immorality, my objection to female suffrage would vanish. Everyone knows perfectly well that if this amendment is correct, then it applies as much to men as to women and I am surprised that the Honourable Gentleman opposite should have brought it forward to deal only with the class he describes as female persons. Of course, this really shows what absurdities we are attempting to legislate upon.

***

There was nothing new in dispatching unruly and 'difficult' girls and women to institutions. Nineteenth-century Britain was filled with receptacles for females who had 'fallen' – reformatories, Anglican 'convent' penitentiaries, secular penitentiaries (such as Urania Cottage at Shepherd's Bush, West London, co-founded by Charles Dickens), industrial schools, homes for inebriates, the workhouse, the prison. Once a sentence in a prison, industrial school or inebriates home was completed, inmates were free to leave; and none of the other institutions mentioned here had power of detention beyond the age of sixteen. In 1886, the Idiots Act had created a specialist set of 'idiot asylums', or 'idiot homes', as the state, for the first time, recognised the different requirements of the mentally ill and the learning-disabled; the most clearly incapable girls and boys could be medically certified into this specialist place of detention. But, as we have

---

* On votes for women, Banbury said in the Commons: 'Women are likely to be affected by gusts and waves of sentiment. Their emotional temperament makes them so liable to it. But those are not the people best fitted in this practical world either to sit in this House . . . or to be entrusted with the immense power which this [franchise] bill gives them' (*Hansard, House of Commons Debates*, Fifth Series, vol. 94, 19 June 1917, col. 1645).

seen, doctors were reluctant to label as 'idiot' or 'imbecile' those with greater capabilities but who nevertheless did not appear to know how to behave, or, for females, how to adapt their behaviour to societal expectations of femininity.

In a case from around 1900, doctors did comply and certified one girl of high intelligence, placing her in an idiot home because she refused, or could not, settle down to the fate that had been decided for her. She had won a prize for a written piece of work in Divinity (Bible Studies) class – yet within months medical men had declared her feeble-minded. 'She was so very tiresome', said the female Local Government Board inspector for the Northern and Midlands counties who oversaw the case. 'She was sent into [domestic] service . . . I am thankful to say the mistresses would have nothing to do with her. They returned her to the [Poor Law] Boarding Out Committee, who could do nothing with her.' At that point the girl was certified as 'mentally deficient' and put away. A Local Government Board colleague had wondered whether the divinity girl was typical of those who have 'exceptional gifts in certain directions and want of gifts in others'. This worrying case history shows that even before the passing of the Mental Deficiency Act, behaviour and 'attitude' were policed, and judgments were made that could lead to a wholly inappropriate certification and detention. What a colossal waste of her immense potential.

Susan Mumm has written on the Church of England's 'female penitentiaries', which flourished between the 1840s and the 1910s; Anglican sisterhoods set up these shelters for a wide range of outcast females. Girls and women were offered penitentiary places either as an alternative to prosecution; or after release from a prison sentence; or, more commonly, as a place that destitute girls could volunteer to enter. The usual length of stay was two years.

Mumm writes that

> penitentiaries were intended as transformative institutions, where female outcasts of many kinds could be changed into 'honest' women, a conversion which incorporated both a spiritual change from sinner to penitent, and an equally important social shift from dissolute and deviant female to respectable woman . . . As well as reforming prostitutes, Anglican penitentiaries in Victorian Britain offered shelter to the survivors of

incest and sexual violence, women fleeing abusive relationships, and female alcoholics.

To twenty-first-century eyes, what is striking is how vanishingly rarely anti-social behaviour or educational underachievement were attributed to childhood trauma, including sexual abuse, violence and neglect. In spite of often detailed personal case histories being taken by philanthropic bodies and Poor Law Guardians, the questions that seem obvious to us were almost never asked. Could this child's seeming 'mental deficiency' or 'moral defectiveness' be attributable to catastrophic experiences in early life? Nobody in officialdom in these years seemed to attempt any causation of this kind – only, occasionally, correlation. (After the First World War, and the gradual acceptance of the discipline of psychology in Britain, we will see this start to change.)

The penitentiary records that Mumm consulted reveal epic levels of child sexual assault, incest, rape, cast-off mistresses, rescued attempted suicides, early-adolescent prostitutes. 'Not all sexually transgressive penitents had engaged in sexual activity voluntarily', continues Mumm:

> The youngest 'adult' penitent in the casebooks studied was thirteen; she had been incestuously abused; the specialised houses for child victims and children 'in moral danger' which several sisterhoods opened in the late-Victorian period accepted children as young as eight . . . There are accounts of both general labourers and wealthy gentlemen committing incest on their female children, thus starting them on the road which led to the convent penitentiary. The casebooks of the communities also indicate that incest within the Victorian family was a frequent precursor of homelessness, which in turn led to prostitution.

Furthermore, using a range of euphemisms, the sisterhood described many of their inmates as having 'been wronged', 'deceived', 'led astray' or 'fallen' while working as a domestic servant. In fact, two-thirds of the entrants to the penitentiaries had been 'undone' while working as a maid.

The Workhouse Girls' Aid Society recognised this phenomenon, too. The Society formed in London in 1880 to give refuge to single pregnant women; they discovered that servants were a significant proportion of their intake. One secretary of the Society said, 'Every girl, however poor, lowborn, defective, unprotected, is precious in the sight of God. It is no kindness to pass over a man's sin lightly because he is a man and to place all the shame and stigma on a woman's sin because she is a woman.' That said, like most philanthropic organisations, the Workhouse Girls' Aid Society tended to be harsher when a girl came in pregnant with her second illegitimate baby.

Domestic service was, then, a potentially dangerous job for girls to enter. But it remained the destination to which most organisations – whether religious or secular – sent those in their care. This was wholly realistic, of course – the range of jobs for unskilled and uneducated women was limited. But it also had the effect of consolidating the lack of opportunities and low aspirations for working-class women – submission rather than ambition, usefulness to others rather than self-determination. An unfair, anachronistic judgment on my part? Perhaps. In their defence, all of these Victorian institutions (no matter how grim, how moralistic, how lacking in a grander vision) admitted of the fact that change was possible for an individual – reform, transformation, improvement, salvation even. What was coming down the line was a 'solution' that labelled and damned certain girls and boys, women and men, as incapable of change.

PART 2

CHANGING THE LAW

6

'BE VERY CAREFUL': WINSTON WADES IN

A bucket of cold water was thrown onto the degeneration pessimists in 1904, with the publication of a governmental report. The Inter-Departmental Committee on Physical Deterioration had been convened in the immediate aftermath of the disastrous attempt to recruit British soldiers to fight the Second Boer War, of 1899 to 1902. Two-thirds of men who had wanted to enlist had had to be turned away because of their poor physical condition – or so it had been reported: later investigation put the figure closer to one-third. (In Germany, the army rejection rate was sixteen per cent.)

The Inter-Departmental report stated that 'so far as the Committee are in a position to judge, the influence of heredity in the form of the transmission of any direct taint is not a considerable factor in the production of degenerates'. All six medical specialists agreed that inheritance played either no role, or a minor role, in the physical weakness revealed in the Army recruitment crisis, with only alcoholism and syphilis possibly having a (limited) transmissibility. Instead, they found that the shattered health of the urban British population was caused by poor housing and labour conditions; urbanisation and overcrowding; lack of good-quality food and clean water; and air pollution.

The Inter-Departmental Committee had been, as its full name implied, concerned only with physical, and not mental, degeneration. Sixty-eight individuals (fourteen of them women) gave their expert testimony, including physicians, surgeons and dentists; educationalists and factory inspectors; philanthropists; even two anthropologists. The rare mentions during the hearings of feeble-mindedness included a devastating swipe at Professor Karl Pearson, demolishing Pearson's statement (see p. 31) about the presumed declining levels of Britons' intelligence. Professor Daniel Cunningham said:

> I think that the statement [by Pearson] is a pure assumption. I do not know how we can possibly measure this supposed loss of inherited intelligence (we are dealing with 'inherited intelligence' because all his remarks refer to inherited intelligence) and I do not think there is a single solid fact in support of such a view. I am astonished that one for whom I entertain so high an admiration as Professor Pearson should have put forward such a statement ... It should be borne in mind that it is stocks [bloodlines] and not classes which breed men of intellect. These intellectual stocks are found in all classes, high and low. No class can claim intellect as its special perquisite.

The report also firmly underlined the fact that no satisfactory statistical surveys were available to work out whether the nation's health was improving, holding steady, or worsening, and that comparisons with the past, and between urban and rural populations, were virtually impossible:

> Disjointed and partial inquiries have taken place from time to time ... but these inquiries were not instituted in any relation to each other, nor conducted on similar lines, so that the results obtained are of very little use for the purpose of comparison.

In addition to the Inter-Departmental/Boer War fiasco inquiry, in the postwar months many MPs had petitioned the government to inquire into the unsatisfactory provision made for the 'feeble-minded'. As shown in chapter 1, the penal, educational, 'lunacy' and Poor Law systems were failing to deal with this nebulous category of individual. What even was 'feeble-minded'? As John Owen has written, in his pioneering study of the Royal Commission on the Care and Control of the Feeble-Minded, 1904–8, the aim of this Royal Commission was primarily to settle an administrative problem of how to deal with the feeble-minded – rather than scientifically to categorise the various grades of mental defect, or to settle on its cause. Nevertheless, this Commission, chaired by the Earl of Radnor (and known from now on in this work as the Radnor Commission), did discuss causation and categorisation, as the environmentalists and the hereditarians once again locked horns. Of the 209 witnesses whose opinion was sought, thirty-five mentioned heredity; and of these, twenty-five attached great importance to a history of mental defect in the family. Three recommended sterilisation of the feeble-minded to prevent them breeding.

The final report of the Radnor Commission estimated that there were about 100,000 mentally defective adults in Britain (excluding certified lunatics) and 48,000 defective children on schools' registers, of whom 12,000 were in special school provision. With the 1901 Census putting the total population of England and Wales at 32.5 million, the defective were found to constitute 0.46 per cent.

After perusing the seven volumes of evidence, it is a surprise to see the final report come down so heavily in favour of the inherited nature of feeble-mindedness. The extremely good environmentalist arguments put forward by Dr Alfred Eichholz, and by forensic psychiatrist Dr Charles Mercier and Dr Robert Hutchinson of the Great Ormond Street Hospital for Sick Children (who threw everyone off guard with his claim that feeble-mindedness was a totally random outcome) were cast aside without serious, detailed rebuttal. On the one hand, the Radnor Commission report continuously bemoans the paucity and quality of statistical data; but on the other hand, it wasn't about to let that undermine its finding that feeble-mindedness was overwhelmingly an inherited condition.

Not even the most brilliant scientist was yet able to explain the biochemical processes by which transmission occurred; and no blame should be attached to hereditarians for that reason. But the lack of caution and the lack of humility in hereditarian arguments are blameworthy, when the consequence would be detention for life for those who were certified as mentally defective.

John Owen has observed that 'the adoption of heredity as the major influential factor in the causation of mental defect by the Royal

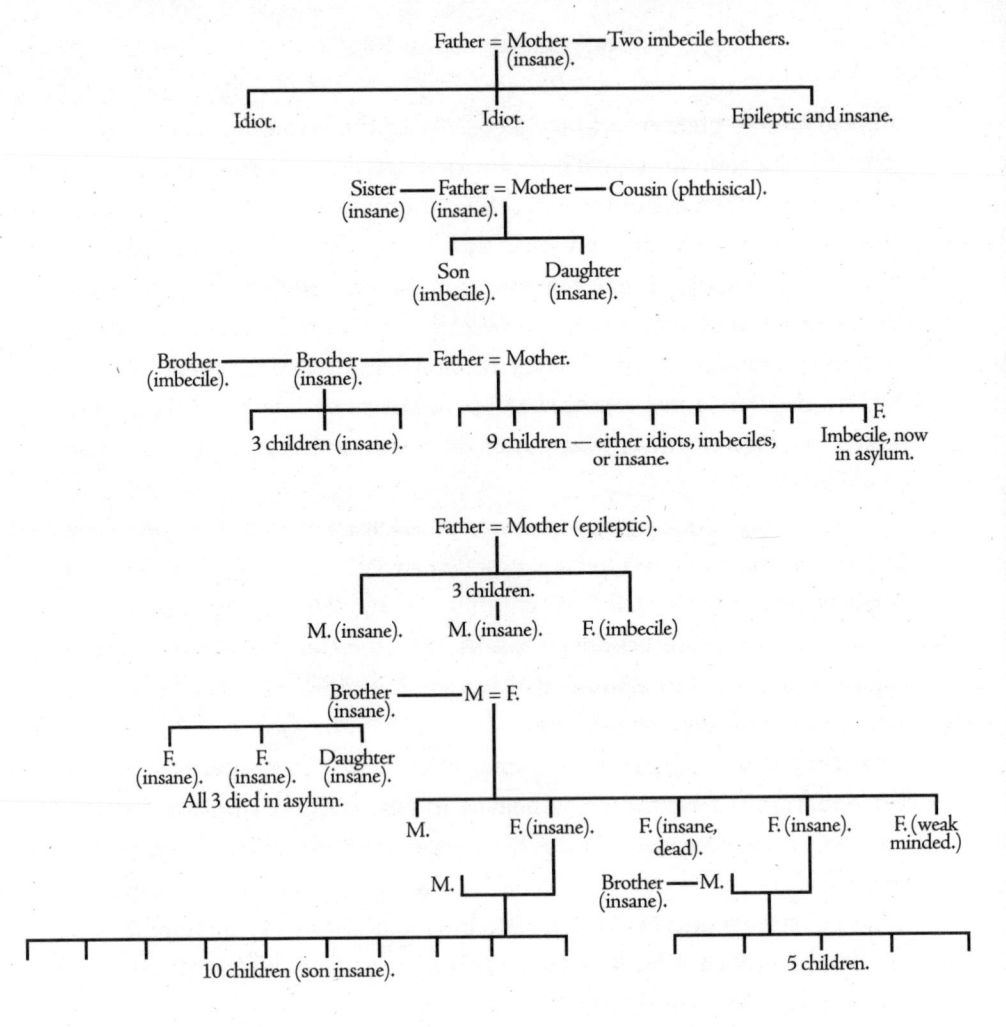

```
                    Father = Mother ——— Two imbecile brothers.
                           │ (insane).
         ┌─────────────────┼─────────────────────────┐
       Idiot.            Idiot.                   Epileptic and insane.

              Sister ——— Father = Mother ——— Cousin (phthisical).
             (insane)    (insane). │
                         ┌─────────┴─────────┐
                        Son               Daughter
                      (imbecile).         (insane).

  Brother ——— Brother ——— Father = Mother.
 (imbecile).  (insane).          │                                    ┌ F.
              │         ┌────────┼────────┬────────┬────────┐  Imbecile, now
         3 children (insane).  9 children — either idiots, imbeciles,  in asylum.
                                          or insane.

                    Father = Mother (epileptic).
                           │
                      3 children.
              ┌────────────┼────────────┐
           M. (insane).  M. (insane).  F. (imbecile)

                 Brother ——— M = F.
                (insane).      │
          ┌──────┬──────┐
         F.     F.    Daughter
       (insane). (insane). (insane).
        All 3 died in asylum.
                      ┌──────┬──────────────┬──────────────┬──────────────┐
                     M.    F. (insane).  F. (insane,    F. (insane).    F. (weak
                                           dead).                        minded.)
                     │                                    │
                    M.                        Brother ——— M.
                                             (insane).
         ┌───────────────┐                         ┌───────────┐
         10 children (son insane).                 5 children.
```

Family trees were frequently presented by hereditarians attempting to prove that mental deficiency was an inherited condition. This tree was created by Dr P.W. Macdonald, a medical inspector at the Dorset county asylum. He offered it as evidence at the 1908 Royal Commission on the Care and Control of the Feeble Minded.

Commission can and has been seen as justification for acceptance of a "Social Darwinist" ideal behind it. The reason for the inclusion of heredity was, however, administrative.'

The administrative simple 'solution' to the problem of the feeble-minded was to place them into sex-segregated 'colonies', for permanent detention, so that they would receive both 'care' and 'control', and would no longer be able to breed more generations of the mentally unfit. The eugenic argument fitted best with the solution that the Radnor Commission already favoured. To accept, instead, an environmentalist view of feeble-mindedness as a basis for policy would have required a complex, slow, expensive, untried series of societal changes which, while not unthinkable in 1908 (when the Radnor Commission reported), were nevertheless still against the grain, politically.

Other solutions suggested by the eugenicists included a marriage ban for anyone unable to provide a certificate of good physical and mental health. The three witnesses who had put forward sterilisation as an answer all accepted that this was a repugnant idea to the public at large, and that to be effective – to stamp out feeble-mindedness altogether – sterilisation would have to take place on a scale that was at present impossible: the non-feeble-minded siblings of the feeble-minded would also have to be sterilised if the taint was to be eradicated entirely.

These illiberal notions would be satirised by one Oliver W.F. Lodge, of Edgbaston, Birmingham, in a letter to the *Daily Herald* newspaper entitled 'Sterilisation of Eugenic Cranks', in June 1912. Lodge facetiously stated that since parents over the age of forty disproportionately had 'unfit' offspring, everyone over that age should be segregated and sterilised.

> The measure should be classed as non-contentious, and a Liberal government might be expected to give it every facility. In days when it is seriously proposed to abolish love and substitute the principles of the stud farm, the sterilisation of eugenic cranks may easily become a matter of urgency.

The Radnor Commission rejected both sterilisation and a marriage ban. Instead, its report recommended that 'mental defectives' who were in the workhouse, on Out Relief, in penal institutions and in lunatic asylums should come under the auspices of a new body, the Board of Control, with local authorities obliged to identify and bring forward any child or adult defective in their community.

What happened next – was not much at all. To the anger of the hereditarians, the government appeared to have no firm plans to act on the findings of Radnor. Various Whitehall departments came together to discuss the report on 17 December 1908, and the Home Office and the Board of Education argued with each other about how to progress: should any legislation on the feeble-minded be primarily custodial (the Home Office's natural remit), or should it be based on highly specialised educational provision? And when, in any case, would there be time to bring in legislation? The landslide Liberal Party General Election victory of 1906 had led to the introduction of a raft of Bills that gave the state unprecedented powers over the domestic and family life of Britain, as well as interventions into laissez-faire in business and industry. At long last, after over thirty years of agitation from various quarters for Britain to modernise itself on so many fronts, a government was elected with the clout to attempt just that.

Legislation to protect the vulnerable and to provide a basic safety net would include the 1906 Workmen's Compensation Act, the 1908 Children Act, the 1908 Old Age Pensions Act, the 1909 Labour Exchanges Act, and the introduction of National Insurance in 1911. In 1909 the first genuinely redistributive Budget – Lloyd George's so-called 'People's Budget' – was passed, which heavily taxed landowners and introduced 'supertax' for those earning over £5,000 a year. The Majority and Minority Reports of the 1909 Royal Commission on the Poor Laws urged an overhaul of the national welfare system. In 1911, the government passed the Parliament Act, neutering the House of Lords and removing its ability to interfere with any finance Bill, or any other kind of Bill that the Commons had passed three times. George Dangerfield would famously, in 1935, describe these events cumulatively as 'the strange death of Liberal England': 'The old order, the old bland world, was dying fast; and the Parliament Act was its not too premature obituary.'

On top of all this, Irish Home Rule, female suffrage, Welsh church disestablishment and defence spending were jostling for parliamentary time, too. This rammed agenda partially explains the reluctance of the Civil Service and MPs to work towards bringing forward any action on the 'feeble-minded' – which was, in any case, highly likely to be a hard sell to the British public. At this point, two out of three men had the right to vote; and so, as never before, policymakers had to keep an enlarged electorate in mind. Throughout the subsequent departmental and parliamentary discussion, references to the unpalatability (to the population at large) of restricting the freedoms of this section of the community were common.

It was in this landscape that the Eugenics Education Society (EES) formed, in 1907. It is better known by its later name, the Eugenics Society, but the central word is crucial to understanding its key role in subsequent events. The EES launched a hugely energetic, well-targeted and emotive campaign of persuasion – sending to members of parliament and to Civil Service senior staff propaganda material that played upon 'race suicide'/ demographic emergency fears. In keeping with hereditarians' highly tendentious arguments at the Radnor Commission, the EES bombarded policymakers and MPs with harrowing anecdotal evidence of feeble-minded people running amok throughout the nation, causing harm to themselves and to their communities, and copiously breeding the next generations of undesirables. The EES's stated aims were to end 'the present conspiracy of silence that envelopes the subject of birth and parenthood'; 'to raise public opinion on questions of morality'; and 'to strengthen public opinion against unhealthy marriages and a wilful propagation of an unhealthy and suffering race'.

Historian Pauline Mazumdar has traced the personal histories of the founders of the EES, pointing out that many had already cut their campaigning teeth in such organisations as the Charity Organisation Society, the Society for the Study of Inebriety, the National Association for the Care and Protection of the Feeble-Minded, and the Moral Education League. Founded in 1898, the latter was, says Mazumdar, 'a model of social activism ... Its members were the socially responsible, advanced thinkers of their day, progressives who actively worked for the programmes in which they believed.' The National Association for the Care and Protection of the Feeble-Minded and the Moral Education League had undertaken a massive campaign strategy of letter-writing to the nation's Boards of Guardians, education committees and other bodies that oversaw the poorest citizens, garnering points of view and case histories to bolster their eugenical argument.

As noted already, a large number of the EES's 341 earliest members were women, and the core committee members were all drawn from the educated middle class, though notably, not from the business community; and while the EES had many doctors as members, the majority of doctors did not join – and in fact the British Medical Association remained mostly hostile to eugenicists' aims. The Church of England was well represented, and the Dean of St Paul's Cathedral, Reverend William Inge, who was one of the more outspoken members of the EES, in 1911 stated:

> I cannot say I am hopeful about the near future. I am afraid that the urban proletariat may cripple our civilization as it destroyed that of ancient Rome. These degenerates, who have no qualities that confer survival value, will probably live as long as they can by 'robbing hen roosts', as Mr Lloyd-George truthfully describes modern taxation, and will then disappear. Meanwhile we must do what we can, which is not very much.

Liberal economist William Beveridge, one of the founders of the Welfare State, did not join the EES but wrote and spoke in support of it, believing that those who, through mental weakness, could not play a full part in the world of labour, could be made 'the acknowledged dependants of the state, removed from free industry, and maintained adequately in public institutions, but with complete and permanent loss of civil rights – including not only the franchise but civil freedom and fatherhood'. As we will see in the next chapter, a significant portion of the late-Victorian and Edwardian Radical Liberals and Socialists had no patience with elements of the pauper class whom they considered parasitical, and holding back 'respectable', 'decent', hard-working labour from earning its full share of an economy's wealth. People with 'progressive' politics sometimes cited capitalism's preference for a large pool of surplus, casual, unskilled labour as a factor in the 'multiplication of the unfit'; this 'residuum', this 'submerged tenth', this 'lumpenproletariat' undermined the working classes as a whole – or so one strand of leftist thinking went.

Mary Dendy and Ellen Pinsent were indefatigable early EES members – and in addition to writing for journals and newspapers were excellent platform speakers, too. In this way, they conjured up a panic where previously there had only been concern. They incited guardians of the poor and educational staff to petition parliament and then passed this agitation off as outraged public opinion. In 1910, the EES and the National Association announced their intention to draft a parliamentary Bill on the compulsory detention of the feeble-minded, and went as a deputation to parliament to seek an MP to present it. Here they met, at last, a fully sympathetic Home Secretary. He stated that there were

> at least 120,000 or 130,000 feeble-minded persons at large in our midst. These unhappy beings deserved all that could be done for them by a

Christian and scientific civilisation, now that they were in the world. But let it end there, if possible. If, by any arrangement . . . we are able to segregate these people under proper conditions, that their curse dies with them and is not transmitted to future generations . . . we shall have taken upon our shoulders in our own lifetime a work for which those who come after us will owe us a debt of gratitude.

That Home Secretary was Winston Churchill, who had come into the post in February 1910. He took an instant interest in the 'feeble-minded question', and wrote the following memo to the permanent under-secretary at the Home Office, Sir Charles Edward Troup.

I am drawn to this subject in spite of many Parliamentary misgivings. It must be examined. How would you suggest? What body? A departmental committee? A Parliamentary committee? What is known about it in the Home Office? Indiana? Other experiments? What is the best surgical operation? Assuming this to be satisfactory – under what authority? Under what conditions? a) voluntary b) in return for a money payment c) compulsory – by order of the court – the Board of Control d) as a condition of release e) young persons? Let us begin with a memo from the dept on the subject. Next I will sound my colleagues. Of course it is bound to come some day.

Churchill had read an eleven-page booklet, *The Sterilization of Degenerates*, written by Dr Harry C. Sharp, who had been sterilising inmates of the Indiana State Reformatory since 1899. He had performed more than two hundred vasectomies on delinquent teenage boys and young men by the time his opponents managed to get him stopped; but in 1907, Indiana legalised involuntary sterilisation of individuals in institutions who were judged to be mentally 'unfit' (it was the first US state to do so) and the practice re-started, with both males and females, the mentally deficient and the mentally ill, being operated upon. Using a blue pencil, Churchill marked in Sharp's pamphlet the sections about the Indiana legislation, and used this to bolster his pointed questions to Troup.

In his reply, two months later, Troup referred to having discussed the matter with Dr Bryan Donkin, who had sat on the Radnor Commission and was a Home Office medical adviser to the Prison Commission; he had travelled to the United States and spoken to many doctors who deplored the Indiana legislation. Donkin was alarmed at the eager and urgent tone of Churchill's memo, and pointed out that sterilisation had been rejected in Britain by the Radnor Commission, for good reasons. Donkin described Dr Sharp's pamphlet on sterilisation as 'a monument of ignorance and hopeless mental confusion'.

Troup didn't necessarily agree with Donkin, he told Churchill, but he reminded the Home Secretary of the huge problem of 'the present state of feeling' in the nation about such a move. In addition, he wrote, the bulk of Whitehall leaned towards finding an educational solution, with permanent detention reserved for those who demonstrably could not benefit from any kind of schooling. The scalpel was not on Whitehall's wish-list at all.

Perhaps, though, Troup continued, when a new 'Board of Control' was established, 'it will be possible to empower, in general terms, to authorise medical operations which would be for the benefit of the weak minded ... Be very careful', Troup ended, enigmatically, probably signalling that this was a topic upon which being forthright could either damage a promising political career, or could further upset the already fairly fraught relations between the Home Office and the Board of Education on this very subject.

Undeterred, five months later Churchill wrote to Prime Minister Asquith that

> the unnatural and increasingly rapid growth of the feeble-minded and insane classes, coupled as it is with a steady restriction among all the thrifty, energetic and superior stocks, constitutes a national and race danger which it is impossible to exaggerate ... I feel that the source from which the stream of madness is fed should be cut off and sealed up before another year has passed ... a simple surgical operation so the inferior could be permitted freely in the world without causing much inconvenience to others.

To try to persuade his Cabinet colleagues on to his side, Churchill circulated an article by Dr Alfred Tredgold, published in the July 1909 edition of the *Eugenics Review* and entitled 'The Feeble-Minded, A Social Danger'. Churchill and his department set about drafting the 'Feeble-Minded

Control Bill', which was still in consultation when Churchill moved from the Home Office to the Admiralty in October 1911. He maintained his keen interest in the subject, though, and in the final week of July 1912, when the first international Eugenics Conference took place in London, at the Hotel Cecil on the Strand, Churchill was a delegate and a vice-president. The diary of one of Churchill's close friends, Wilfred Scawen Blunt, reveals that in October 1912 the two men had discussed eugenics. 'Winston is also a strong eugenist', wrote Blunt.

> He told us he had himself drafted the Bill which is to give power of shutting up people of weak intellect and so prevent their breeding. He thought it might be arranged to sterilise them. It was possible by the use of Roentgen rays [X-rays], both for men and women, though for women some operation might also be necessary. He thought that if shut up with no prospect of release without it, many would ask to be sterilised as a condition of having their liberty restored. He went on to say that the mentally deficient were as much more prolific than those normally constituted as eight to five. Without something of the sort, the race must decay. It was rapidly decaying, but could be stopped by such means.

Churchill's enthusiasm for the second-most extreme form of eugenic interference is an embarrassment to those who idolise him. For those who are deeply sceptical of Churchill when the entirety of his political career is considered, his eugenics is quite in keeping with his disdain for the non-white races and for many, perhaps most, of the working class.

As Home Secretary, Churchill had urged caution regarding the 'preventive detention' of 'habitual criminals'. This illiberal measure had, as already stated, been introduced in the 1908 Prevention of Crime Act. He worried about the definition of 'habitual' and about a likely absence of standard procedure across the nation's 180 police forces and the lack of expertise on the topic on the part of the judiciary. He pushed for the improvement of the British prison system and for the ending of imprisonment and fines for both young offenders and petty offenders: he did not want such people as repeat pilferers, for example, or vandals, being detained on the mere say-so of the Director of Public Prosecutions. However, this enlightened attitude is nowhere to be seen when, in the very same month, he called for 'tramps and wastrels' to be dispatched to 'proper labour colonies, where they could be sent for considerable periods and made to realise their duty to the State'.

His devotees point out that Churchill repeatedly argued that sterilisation of the feeble-minded would be a humane measure, in that it would allow them to avoid a lifetime of segregation in a 'mental deficiency colony', as recommended by the Radnor Commission. Instead, neutered, they would have their freedom within the community – now that they were unable to propagate their kind. Whether or not this was a sincerely held stance, it is clear that he had begun to think in biological terms before the Liberal Party developed its plans to overhaul welfare provision: 'The improvement of the British breed is my aim in life', Churchill had written to his cousin Ivor Guest, shortly after his twenty-fifth birthday in January 1899.

When Churchill joined the Liberal Party from the Conservatives, in 1904, he fully embraced 'New Liberalism' and its ambitious programme of attempted systemic change to create both greater 'social efficiency' and higher economic output by improving the living and working conditions of the British people. However, he showed insensitivity to the realities of working-class lives in an exchange in the House of Commons on 23 February 1911 regarding 'The Epping Case', in which Mrs Annie Woolmore had been imprisoned in Holloway for six weeks for neglect of her five children, despite a local doctor testifying that this was a case of chronic poverty, not of parental neglect. James O'Grady, Labour MP for East Leeds, asked Churchill if he would intervene to obtain a review of the sentence, since the five Woolmore children were now pining in the workhouse at Ongar, and Mrs Woolmore was ill in prison. O'Grady said she was known to be of sober habits, had recently been in very poor health, and because there was no piped water in her rundown rented cottage, she had to walk a mile with pails to obtain water, with her youngest child on her back. The children were reported to have looked unclean but had been found to be in good health. In his reply, Churchill said that O'Grady's report was inaccurate – that Mrs Woolmore was 'weak minded' and kept the cottage and the children in 'a filthy condition'; that she had a pond very close by for water to wash with, and drinking water was no further than 200 yards away. It was in the children's interest that she stayed in prison, he said, and he hoped the NSPCC would make some arrangement for the family upon her release.

O'Grady repeated that the only matters of alarm had been the 'insanitary' rented cottage (actually a wooden shanty, and allegedly owned by a member of parliament), and the fact that the Poor Law allowed Mrs Woolmore just thirteen shillings and sixpence a week for herself and the children; her own poor health and mental confusion, O'Grady said, had been brought about by starving herself to be able to feed the children. 'The whole family life has been decimated', he told Churchill, so shouldn't the woman be entitled to some reconsideration of her sentence? Could she spend time in an infirmary to regain her strength? But Churchill would not budge.

Local Epping feeling, outrage in a handful of national newspapers and, possibly more importantly, the intervention of high-profile women's rights campaigners Lady Constance Lytton and Charlotte Despard saw 'The Epping Case' bubbling up into a cause célèbre; and the Men's Committee of the Justice for Women Association placed on record its 'profound regret that the occurrence of such a flagrant injustice and harsh treatment should be possible in a civilised country'. In the *Daily News*, G.K. Chesterton asked 'Are We All Mad?' in an extended column on Mrs Woolmore's situation. 'I challenge any person . . . to tell me what that woman was sent to prison for. Either it was for being poor, or it was for being ill.' Churchill then relented and Mrs Woolmore was released from Holloway after four weeks. There was not, and had never been, any suspicion that she was 'feeble-minded' – Churchill had simply made up his own mind that she must have been so, failing, or perhaps refusing, to make the connection between starvation, struggle, illness and mental fatigue.

***

The Eugenics Education Society and the National Association's own Feeble-Minded Persons (Control) Bill was taken up by a cross-party parliamentary committee, and on 17 May 1912 it was presented to parliament by Conservative and Unionist MP for Wirral, Gershom Stewart. He told the Commons that 'the object of this Bill is to regularise the lives, and, if possible, to prevent the increasing propagation of half-witted people.' It passed its second reading unopposed, before being withdrawn once Churchill's replacement as Home Secretary, Reginald McKenna, announced that the government now had ready its own version of such legislation, the Mental Deficiency Bill.

This, too, easily passed its second reading, in July 1912, but was lost in committee; even though it had a healthy majority in the Commons of 230 to 38. A particular issue at committee stage had been the clause criminalising marriage with a defective, 'or to solemnise, procure or connive at such a marriage'. This was clearly going too far, even for those supporting the Bill. In addition, at committee stage, four months of discussion had failed to define 'defective'.

So Home Secretary McKenna then introduced his second, revised Mental Deficiency Bill, with all overt references to eugenics now stripped out (or, rather, secreted within highly euphemistic language). This Bill sailed through the Commons and the Lords, receiving Royal Assent on 15 August 1913, to come into effect seven months later.

The parliamentary movers of all three Bills had ventriloquised the EES and the National Association, presented the poorly evidenced hereditarian opinion as irrefutable fact, and characterised fastidiously cherry-picked vox pops as a vast wave of popular opinion. Those in favour of permanent sex-segregated incarceration of the feeble-minded frequently mentioned how very kind such legislation was – it was, in fact, the moral option, to shut them away. Charles Scott Dickson, Conservative-Unionist MP for Glasgow Central, claimed that 'in America I have seen several institutions, and my firm conviction is that the great majority of the patients are so comfortable and happy that they do not want to go away'.

Gershom Stewart quite agreed that 'what we really wish to do is to free the sufferers as far as we can from the bondage of their own defects'. Home Secretary McKenna pointed out what a burden freedom actually was for these people, claiming, 'To a large class of the feeble minded, the greatest misery to them is the responsibility of liberty ... they are happy only when the sense of personal responsibility is taken from them.'

And just think of the taxpayer! The prevailing chaotic system of housing the feeble-minded across special schools, workhouses, penal institutions and lunatic asylums was costly, in comparison to the proposed specialised, purpose-built system of 'colonies' where they could be concentrated together and prevented from breeding. Referring to the Radnor Commission, Gershom Stewart spoke of this sort of women [sic] coming in year after year and producing children from different fathers and from fathers perhaps whose names they do not even know . . . Those children the ordinary taxpayer has got to carry from the day they are born until the day they die.

The new colony regime would be a far more efficient use of public money, agreed Sir John Rees, Liberal Unionist MP for Nottingham East: 'The procedure provided is as lenient as may be to the taxpayer, which I continue to think is one of the greatest possible benefits in any Bill brought before this House.'

Embattled opponents of the Mental Deficiency Bills claimed that the legislation would remove parental say-so on the future of their troubled child – that the state would place itself firmly into the centre of family life, in a way not seen before in British society. These were the same arguments that had, ultimately unsuccessfully, held up the introduction of child abuse and neglect legislation, as well as increased legal protection for mothers and wives in an abusive domestic set-up. (Measures of this kind began to be passed from the 1850s onwards.)

Perhaps even worse, such a draconian new Act would lead to the repeat of the abuses seen across the nineteenth-century lunacy system, namely, malicious or mistaken asylum detention, and the neglect and abuse of patients who had fallen off the radar of the asylum inspectorate. Why wouldn't these phenomena occur in a mental deficiency colony system? Where were the safeguards?

Willoughby Dickinson, Liberal MP for St Pancras North, and a seconder of the original attempt at legislation, pledged that only when it was absolutely certain that a person was feeble-minded and a source of danger to themselves or to the public would they 'be put into these homes by forcible detention, and, in my opinion, these cases will not be very numerous . . . [they can] be detained with sufficient and proper safeguards'. Opponents would just have to take it on trust that the finer details about ensuring no improper detention would be implemented after the Commons and Lords had approved the Bill. This scarcely answered the question.

7

THE MAN WHO SPOKE DRIVEL

He was 'the obstructionist in chief', 'the handsome young Liberal-Anarchist [who] attacks his own government in and out of season'. Forty-year-old Josiah Wedgwood would, in the end, be the only serious parliamentary opponent of the mental deficiency legislation. By the early 1900s Wedgwood was something of a political anomaly: a left-leaning libertarian, who backed progressive measures on the part of the state that would promote the rights of workers, women and the colonised people of the world, right up to the point that such measures came into conflict with personal, individual liberty. His belief was that the interests of the state must always come second to the freedom of the individual.

His party, the Liberals, were becoming increasingly split on this very topic. The fissure was becoming ever deeper from the 1890s, as many adopted the 'New Liberalism' – a rejection of laissez-faire and the recognition that a degree of collectivism was crucial in order to boost the efficiency of the British worker. For Wedgwood, though, when collectivism clashed with liberty, he would not budge.

A descendant of the Potteries ceramics dynasty, by profession Wedgwood was a naval architect. His niece describes him, in her affectionate biography, as having 'a charm which the years developed into one of his surest weapons. Indignant dowagers and crusted colonels who got into his company somehow by error, thinking him little less than Antichrist, thawed into genial streams of small talk under his sunshine.'

Wedgwood was a teetotaller, and kept himself going during the Mental Deficiency Bill debates with bars of chocolate (which sympathetic

Colonel Josiah Wedgwood, last man standing in the fight against the Mental Deficiency Bills.

Conservative MP Gerald Hohler brought into the House for him) as he put down close to 200 amendments, losing every single vote. Towards the end, he fought from 3.45 p.m. on Wednesday 28 July 1913 to 4 a.m. on Thursday 29th, and then again from 11 p.m. that Thursday through to 5 a.m. on Friday 30th. This set a record for parliamentary obstruction. Nobody found out until much later that his wife of twenty years had just left him, explaining that she had fallen out of love. 'I was nearly off my head at the time', he recalled in his memoir of 1940. 'Black night had come upon me. I did not know where my wife was. Everything I did was to stop myself thinking . . . So I do not like, even now, to look back on that MD Bill.'

His parliamentary supporters on the Mental Deficiency Bill(s) deserted him, too; during the first debates, Wedgwood mustered thirty-eight MPs to vote against the measures, and by the end of the battle, they numbered just two. On the face of it, this lack of backing is surprising, since Wedgwood's arguments were far from controversial. They would have been standard parliamentary fare any time in the previous century-and-a-half. At the most basic level – he pointed out – no generally accepted definition of 'feeble-minded' had been settled upon. Everyone from the severely incapacitated right through to the 'slightly abnormal' individual who exhibited somewhat eccentric behaviour came under the umbrella term, Wedgwood stated. And fundamentally, he asked, why should the nature of British governance undergo a sudden change, with decisions on personal freedom being granted to 'specialists', who would take upon themselves the sinister powers that in former, unenlightened eras had been exercised by emperors, kings, autocrats, priests and witchfinder-generals? Wedgwood was troubled by the inability of the average citizen, or even the bureaucrat, to mount any argument against 'expert opinion'. Experts changed their minds all the time, he argued, so why pass such a Bill when in, say, a decade, the specialists may be thinking very differently about mental incapacity?

> It is this government by specialists that you cannot argue with. Here you are alienating these powers, powers which were possessed by the Holy Inquisition, and you are going to entrust them to a body of specialists, whose absolute remedies for disease change every year . . . By this Bill a large part of the population of this country is handed over to the specialists without any right of taking action at law against those concerned if they put you in prison without any semblance of justice . . . It reminds me in a way of the smelling out on the East African coast by the witch doctor.

That the power of lifelong detention should be handed to psychologists and educationalists seemed to Wedgwood a worrying change. Such fears weren't that new, though, as Wedgwood would surely have known. The 'specialist', the 'expert', had been a bogeyman throughout the second half of the nineteenth century with regard to the 'mad doctors'; and even the head of the Lunacy Commission himself, Lord Shaftesbury, was outspoken about the threat he believed they posed. He told the 1877 Select Committee on Lunacy Law: 'You may depend upon this: if ever you have special doctors, they will shut up people by the score.'

Wedgwood kindled a particular contempt for the experts of the Eugenics Education Society – describing them in parliament as 'eugenic cranks'. His idealism sounded a little old-fashioned (for 1912/13) as he told the Commons that

> the spirit at the back of the Bill is not the spirit of charity, not the spirit of the love of mankind. It is a spirit of the horrible Eugenic Society [sic] which is setting out to breed up the working classes as though they were cattle. The one object in life of the society seems to be to make mankind as perfect as poultry. But we are not ants, bees, and wasps; we are human beings, and all this form of . . . eugenics seems to me to be the most gross materialism that has ever been imported into human society. I think those people who are so anxious to improve the breed of the working classes had better remember that there is such a thing as a soul and that the mere desire to turn people into better money-making machines is merely some horrible nightmare of H.G. Wells.

(An unfortunate reference, as Wells at this stage was a eugenics enthusiast, as we have seen.)

The newspaper press criticised his obstructionism. 'Men of ordinary intelligence cannot be expected to stand the sort of drivel that is now being talked in the House of Commons. In order to delay progress with the Mental Deficiency Bill, of which they seem to be in terrible dread, a small band of radicals have entered the lists as deliberate obstructionists.' The unimpressible *Newcastle Daily Chronicle* continued in this vein, attacking the MP for St Pancras East (Mr Martin) and his 'latest fit of flatulence'. Moreover – where was everyone? 'At no time could you see more than forty members, and they listened as silently and stolidly as though they had all belonged to the peerage.'

***

The parliamentary debates on the feeble-minded reveal one worldview ceding to another. In one world, the liberty of the individual was (either honestly or speciously) held to be the bedrock of British life – a parliamentarian could cite liberty largely without fear of ridicule; in the new world, government, informed by science and sociology, should be permitted to restrict a section of the population in order that the rest could thrive, industry could boom and an Empire be protected. It was no longer good enough to talk 'drivel' about freedom – as Wedgwood had been accused, by the head of the NSPCC, no less. Liberal MP for Northampton Charles McCurdy was similarly scathing of Wedgwood's version of freedom, stating that 'it is not by building up temples of liberty and by polishing and repolishing political constitutions that the people can grow up to proper manhood and healthy national life . . . The new century has brought new social problems, and it is with one of these problems that this [Mental Deficiency] Bill makes a courageous effort to deal.'

Liberal MP Frederick Cawley said Wedgwood's abstract concept of personal freedom had little meaning when applied to material reality – it was 'the kind of liberty that at present is represented by being in and out of the [workhouse] casual wards, the maternity wards and the prisons'. That was 'licence', not 'liberty', said Cawley.

For her part, Mary Dendy of the National Association for the Care and Control of the Feeble-Minded addressed Wedgwood directly in the letters page of *The Times* in early June 1913, stating that the libertarian position he held was endangering vulnerable children, who needed to be removed from abuse and neglect by their feeble-minded family members:

> It is a curious perversion of the desire for justice and liberty which would leave helpless little children at the mercy of human beings who are themselves the slaves of their animal passions. It is in the name of liberty that young girls are brutally ill used, little children murdered, outrages of all kinds committed; and we are asked to allow this to go on indefinitely! So far the publicity given to Mr Wedgwood's wild assertions has resulted in making it more difficult to protect the few children whom we have been able to reach. It is an achievement of which we may be proud. I think he would not be if he could understand the harm he is doing and has already done.

This was unfair: Wedgwood did believe that at-risk children should be protected within institutional settings, and in fact admired Dendy's Sandlebridge homes (though he did not agree with her regarding *permanent* detention). In opposition to government plans, Wedgwood wanted to see the creation of a much-improved voluntary sector, just like Dendy's, for those who really did need residential care. (And for what it's worth, the NSPCC found that feeble-minded parents were *less* likely to mistreat their children than 'normal' parents. Dendy was wrong on this point.)

Labour MP Will Crooks went furthest in the parliamentary debates, using extreme language against the 'feeble-minded', who he claimed did nothing but hold back the 'ordinary' working class. It's an extraordinary outburst, worth quoting at length:

> We hear of liberty of the subject. Liberty of what? Is it liberty to corrupt and pollute the town and countryside by every vicious man and woman who goes there . . . I sometimes feel inclined to prosecute the fathers and mothers of those children for not allowing them to be cared for, and instead of that allowing them to be pushed and chased about the streets . . . To talk about permanent imprisonment is all rubbish . . . I have taken part, in discussions in this House on the unemployable; I have taken Members in authority on both sides and shown them 300 or 400 men not one of whom would be privately employed by any person for anything at all, not even for their keep. These were formerly mentally defective children who had been allowed to drift about the world and to become absolutely useless. There is only one fitting description; they are almost like human vermin. They crawl about, doing absolutely nothing, except polluting and corrupting everything they touch. We talk about the liberty of the subject. What nonsense! What waste of words! We ask that you should take these people and have proper control over them, because they have no control over themselves. They are verminous, dirty, with no idea of washing or cleansing themselves. Yet they are human beings, and you could, under proper control, so far improve them that they could be put to some employment, not enough to keep them (I never expect that) but sufficient to maintain themselves partly, and to give them a human existence which they have not got now. Above everything else, you would stop the supply of these children – a very important thing.

This hatred, expressed by a genuine member of the working class (Crooks was born poor in Poplar, East London, spent time as a child in the workhouse, was self-educated and a passionate trades union activist, and became Labour's fourth member of parliament), is one example of how

Will Crooks, Labour MP, was keen to see the Mental Deficiency Bills passed. Crooks told the House of Commons that it was 'very important . . . to stop the supply of these children'.

split the left was on the topic of mental deficiency. The Fabians, also, were overwhelmingly supportive of harsh measures against the seemingly incapable and inefficient; such a policy sat well with the Fabians' technocratic programme to tackle chronic poverty and industrial and social 'inefficiency'. Middle-class technocrats themselves, the Fabians expected to be in the driving seat of the new welfarism – it was not to be left to the working classes themselves. Fabian Sidney Webb wrote that what was required to improve the condition of working people, and so strengthen British industrial output, was an end to 'the breeding of degenerate hordes of a demoralized "residuum" unfit for social life', as he put it.

Wedgwood actually described himself as a 'socialist' and would go on to join the Independent Labour Party in 1919, even briefly having a role in Ramsay MacDonald's Labour Cabinet in 1924 (as Chancellor of the Duchy of Lancaster). Wedgwood believed in trade unionism; he thought that collective bargaining, paradoxically, would unleash the potential for greater individualism for the worker. But he refused to think of his fellow Briton primarily as a unit of labour – they were instead 'souls', 'humans'.

Will Crooks et al. believed the working class needed to be rid of the undesirable element – the feeble-minded. Wedgwood, by contrast, saw the Mental Deficiency Bills as legislation very much against working-class interests. He pointed out in parliament that it was only the less well-off who would fall within the purview of the legislation, since the wealthy would continue to shield their own learning-disabled children and family members with private, home-based provision. With the new law, teachers of working-class children would be obliged to report any who appeared dull and to push them forward for examination with a view to certification as mentally defective. More than this, he worried that working-class parents would be branded 'unreasonable' if they withheld consent for their children to be certified and put in a colony, or placed under 'guardianship'.

The authorities were also to be given power to judge the quality of a family's living conditions, and this would almost certainly involve making class-based judgments. Wedgwood cited the figure of the Poor Law 'relieving officer' – the man who made welfare decisions in his locality.

> The relieving officer is already a terror in the poorer classes of society; he is the most important person in the poor streets. In places, the relieving officer is more important than the policeman. He is the man to whom you are giving this terrible, this tremendous, authority over all these poor people throughout the length and breadth of the country.

Along with the school authorities, the relieving officer might just seem, in working-class neighbourhoods, a dangerously out-of-sympathy authoritarian figure – seeking out infants to report up the line for ascertainment as mentally defective. 'I do think it shows an absolute want of understanding of the ordinary normal feelings of the ordinary poor people of this country to imagine that they will tolerate such a Bill as this, or such a monstrous injustice as that people are to be sent to prison for life for merely being abnormal.'

He wanted instead to build 'a new and better society', with long, slow work to bring about 'social regeneration', rather than kneejerk legislation. Social regeneration would be 'the final cure of deficiency', he wrote.

To discuss these ideas, in December 1912 Wedgwood formed the Freedom Defence League. The League sought to promote both personal freedom and truly democratic, popularly responsive government; and to 'combat the encroachment of the bureaucracy'. The make-up of this tiny organisation (there were never more than twenty members) reveals the diversity of libertarian opinion at this time, bringing together, for example, Socialists George Lansbury, H.G. Wells and Russell Smart (a founder member of the British Socialist Party) with social conservatives who included G.K. Chesterton and Hilaire Belloc. And its diversity was fatal, with arguments erupting on day one. Wedgwood recalled wryly that 'it remained active and united for nearly two months'.

Russell Smart's organ, *The Socialist*, a Glasgow-based weekly newspaper, repeatedly attacked the mental deficiency legislation, on grounds that this was an attempt to crush vital aspects of what it meant to be human:

> If these people [eugenicists] are to be permitted to inflict their salacious rubbish on society, then God help the working class. It won't be the ... degenerate, drunken, syphilitic clients of [West End brothel-keeper] Queenie Gerald that will suffer from their inquisition, but the already starved and robbed inventor, artist, poet, musician and rebel of the poverty-stricken class. Study this Mental Deficiency Bill and shout against it at every corner.

With even greater hyperbole, *The Socialist* commented on the possibility that the Act might contain a sterilisation clause:

> It gives the capitalist through his various legislative agencies the right to restrict the population – to 'sterilise the unfit', as they put it – which in plain English implies the right to invade the home of the proletarian, to outrage and defile the sanctuary of man and woman's love, and breed, like a stud farm, a proletariat of docile and serviceable type, and in such numbers as the capitalist requirements call for.

And not one single Labour Party MP had voted against it, *The Socialist* pointed out – not even Keir Hardie, not even George Lansbury, who both abstained. Two Labour Party leaders had actually given support to the measures (Arthur Henderson, leader 1908 to 1910) and George Barnes (leader February 1910 to February 1911). Future leader Ramsay MacDonald voted in favour, too.

At the 1913 Labour Party conference the motion was carried, 'that this conference expresses its approval of the principles of race improvement underlying the Mental Deficiency Bills but affirms its belief that no real advance in eugenic reform can take place until the evil anomalies of our present capitalistic system are abolished and the principle of equality of opportunity permeates all branches of our national life.' (Around the same time, the conference of the National Union of Women Workers of Great Britain and Ireland backed the Mental Deficiency Bill but stated that this support was conditional upon the Bill's wording being amended to be less harsh.)

The left-leaning popular newspaper the *Daily Herald* gave Dr Montague David Eder many column inches in which to argue against the Bill. Eder was a socialist and would go on to be an early adopter of psychoanalysis in Britain – he was arguably the first important psychoanalyst active in this country. Eder asserted that the British public were very unhappy about the pending legislation. He said he was hearing everywhere a disgruntlement with the proposed changes, citing

> public opinion, which was everywhere (outside official circles) perceptible in train or tram or bus . . . and one or two weekly newspapers . . . It is only poor people and poor people's children who will be made to suffer imprisonment, or, as the Bill euphemistically calls it, sent to an institution . . . Any poor wretch convicted of stealing a doormat or a loaf of bread can at once be rushed into the institution. The maker of illicit gains on the Stock Market will escape.

For its part, Keir Hardie's newspaper *The Labour Leader* stated that feeble-minded children were 'the outcome of poverty, low wages, bad housing, and the terrible stress of modern life. The House should address itself to the removal of these causes, not tinker with their effects.' Hardie himself said not a word on the entire issue.

With Liberals, Labour and Socialists split on the issue of the mentally deficient, the Roman Catholic Church presented the sole united front. At the turn of the twentieth century there were 1.5 million Roman Catholics in England and Wales – five per cent of the population. The majority were urban and of the working or lower-middle class; and their politics tended to be Liberal, and later Labour. Without the vote, the majority of Catholics did not yet need to be courted by politicians – but that would change as the franchise broadened to include all adults over the age of twenty-one by 1928. Passionate convert to Catholicism G.K. Chesterton wrote in his book *Eugenics and Other Evils* that ' "feeble-mindedness" is a new phrase under which you might segregate anybody'. The Catholic journal *The Month*, in 1912, attacked the 'gross materialism' of which Wedgwood spoke, using these words: 'Man has ceased to be a creature of God. He is a polymorphic agglomeration of protoplasmic molecules!'

The Roman Catholic Church would remain a constant critic of mental deficiency legislation, and by the late 1920s, most elements of the secular left would join it to denounce both the Act and the eugenics movement, too.

***

In 1914 the government imposed unprecedented restrictions on personal freedoms with the first in a series of Defence of the Realm Acts (DORA). A technologically sophisticated new form of warfare and espionage appeared to warrant such drastic action being rushed through at top speed and with little dissent. DORA even had initial broad support from Josiah Wedgwood, who believed the threat from Germany was a single instance in which the state could legitimately override individual liberty for the common good. But, as he knew better than most, the libertarian argument had already been lost by the time of DORA – with the almost total rout in parliament of his arguments against confining for life people deemed to be feeble-minded.

So what legislation had actually been passed by parliament? The 1913 Mental Deficiency Act recognised four classes of mental defective (see Appendix 1 for greater detail on how the Act was intended to operate): idiots, imbeciles, the feeble-minded (a new category) and 'moral imbeciles' (also new), the latter described as 'persons who from an early age display some permanent mental defect coupled with strong vicious or criminal propensities on which punishment has little or no deterrent effect'. The condition was to be present at birth or developing at an early age, rendering the sufferer unable to 'compete on equal terms with their normal fellows . . . or of managing themselves and their affairs with ordinary prudence'.

Individuals in the four categories could be placed into institutions (the newly created mental deficiency 'colonies') or remain out in the community under guardianship, or statutory or voluntary supervision – three different levels of 'care and control', decided according to the risk

An example of a form committing an individual
under the Mental Deficiency Act, 1913.

assessment of the youngster. A guardian took full control of the defective's circumstances and actions, possessed the legal powers of a father and was obliged to ensure the defective did not access either alcohol or drugs. The Mental Deficiency Act made it explicit that defectives should have no kind of sexual encounter, and the guardian was obliged to ensure such attachments never formed. This was to prevent illegitimate and 'congenitally defective' children being born, as per hereditarian ideology and as per those who saw such offspring as too expensive for the nation to maintain; but it was also to protect young males from the bogeyman figure of the corrupting homosexual, believed to be a danger to youth in any community, whether urban or rural.

The local authority (that is, the council) could place a borderline case – not certifiable as defective but in need of additional oversight and training – either 'under statutory supervision', remaining in the family home but subject to social services visits and reports; or 'under guardianship', with the parent or relative duty-bound to supply care appropriate to a mentally defective person.

The local authority's Mental Deficiency Committee had to pre-approve all guardianships, and no change in guardianship could occur unless the committee had been notified and had granted permission.

Guardians received a maintenance payment for the cost of housing, feeding and clothing a defective. As the years went by, and a shortage of potential suitable guardians became apparent, guardianship was often granted to the 'defective's' own family, with social services overseeing the domestic set-up.

Crucial to the working of the Mental Deficiency Act was the huge voluntary sector, which operated with a government grant and was ultimately answerable to Whitehall. By 1939, there would be a total of forty-eight voluntary associations undertaking mental deficiency work across England and Wales.

In brief, those who could petition for a person to be certified as mentally defective were parents, close relatives and, more rarely, a 'friend' of the family; county council education officers; magistrates and county court judges; a voluntary association; and – unusually – the Secretary of State himself.

If any other third party were the petitioner, they had to state clearly in the accompanying paperwork why they believed they were entitled to petition and what role they played in the alleged defective's life. This was to deter negligent, ill-founded or malicious attempts to declare someone mentally defective.

One of the aspects of the Act that would prove most problematic was the ascertainment of a youngster who was 'abandoned, neglected or without visible means of support' and needed 'a place of safety'. Many who had no intellectual difficulties would find themselves scooped up into the mental deficiency system, more or less because they were poor and had nowhere else to be placed.

One year after certification, the defective's case would be reassessed to see whether they should remain in the institution/under guardianship. Thereafter, their case was re-examined at five-yearly intervals. Upon reaching the age of twenty-one, all defectives were to be reassessed to see whether they could live independently and earn their own living and so be released from their certification.*

The 1913 Mental Deficiency Act's clause 2 (vi) created another new sub-category – and this was a real shocker: any female 'in receipt of Poor Relief at the time of giving birth to an illegitimate child or when pregnant of such child' was to be ascertained for mental deficiency. As Josiah Wedgwood had tried to warn, this made very poor women and girls particularly vulnerable to mis-categorisation as mentally defective, since the thinking behind this clause was that her pregnancy indicated that she was unable to protect herself (either through weak-mindedness, or her 'erotic tendencies', to use Alfred Tredgold's terminology). Her pauperism singled her out from other unwed mothers. This was a new and nasty alteration to the civil and legal position of the impoverished pregnant unmarried female – to be targeted for ascertainment and institutionalisation, with her baby taken from her. Need it be said, the putative father of the baby rarely faced sanction.

Josiah Wedgwood had noted in all the parliamentary debates how often the promoters of the mental deficiency legislation focused upon unwed mothers. 'This Bill is principally applied to women', he had said. 'No!' roared various members, interrupting him. 'Well', he responded, 'every argument brought forward has dealt with those unfortunate women who go into our workhouses to have children and so on, and the main argument has been against women and against poor women only . . . It will bring her within the meshes of the law . . . This Bill is eminently a Bill which we, as men, have no right to pass.'

He was heavily defeated on these clauses, garnering no more than two supporters. The Commons instead agreed with Home Secretary McKenna, who had argued that if there was any class that ought to be dealt with it was 'the considerable numbers of feeble-minded mothers who go into the workhouses regularly for a year or two, leaving their children behind them, and then go out again into the world'. Willoughby Dickinson MP had chipped in with his observation that 'in many cases, these women were not capable of being mothers and had not the ordinary maternal instinct of an animal'.

---

* Greater detail on how the Act was intended to work is in Appendix 1, on p. 265.

Josiah Wedgwood (left) and Home Secretary McKenna
battled each other for hours in the House of Commons.

Was it getting pregnant while unwed that proved a female was a mental defective of the moral imbecile category – that is to say, was her conduct all the evidence that was needed for such a diagnosis? Or was she already a defective who was then either 'taken advantage of' by a male, or who gave in to her own passions? This was never fully clarified and was left to those in decision-making roles to interpret as they saw fit. The possibilities for the misuse of this clause are clear: a pregnant female pauper, obtaining Poor Relief, was now vulnerable to having her pregnancy interpreted as proof that she was feeble-minded, and ought therefore to be either institutionalised or placed under guardianship, without her child. Poverty and 'social failure' for the first time became bound up in a medico-legal definition and process.

The moral imbecile category would always be the most controversial of the four mental deficiency categories in the 1913 Act. As historian Mathew Thomson has pointed out, it would be the least used of the four; but, Thomson rightly adds, moral assessment also more generally formed part of the certification process of *all* types of mental deficiency, not just moral imbecility. 'Since individual control over moral behaviour was believed to be a direct reflection of temperamental and therefore mental development, persistent moral misdemeanour was regarded as a likely sign of mental defect', Thomson writes.

Dissatisfaction with the moral imbecile category rested partly on the unsolved conundrum that habitual serious criminal or anti-social behaviour did not relate in a straightforward way to intelligence or intellectual ability. As noted in chapter 2, the psychopath, for example (the very worst grade of 'moral imbecile'), tended to have at least average, and sometimes high, intelligence. For his part, Alfred Tredgold, to whom everyone listened with attention (even if they did not ultimately share his views), said that 'absence of remorse always distinguishes the moral imbecile' (this is also a key part of our modern-day diagnosis of psychopathy). But Tredgold admitted that the relationship between intelligence and immorality was 'a question which it is very difficult to decide'. This undecided 'question' made an incoherent mess of many of the mental deficiency incarcerations, a fiasco compounded by the newly created Board of Control's diminished answerability and accountability; the Board now replaced the Commissioners in Lunacy, who had had oversight of the nation's mentally ill and mentally disabled since 1845. Had any of the Victorian anti-lunacy-law campaigners lived to see this turn of events, they would have been plunged into despair.

The Mental Deficiency Act came into force on April Fools' Day 1914, like a bad joke. Challenges to lifelong incarceration on moral imbecility grounds came quickly; but, as we shall see, not quickly enough for thousands of youngsters.

PART 3

'DEFECTS OF CHARACTER AND TEMPERAMENT'

8

'A CROSS BETWEEN PUBLIC SCHOOLS AND GUILD TRAINING COLLEGES': THE MENTAL DEFICIENCY COLONY

'Goodbye, lad. We hope you will soon settle in', said the men from the council Mental Deficiency Committee, showing David Barron into a large dining hall with, David guessed, close to five hundred people seated. Jostling for precedence in David's memory were bars on the windows and the clatter and clang of keys locking and unlocking each room and corridor.

It was the late 1930s, and David was in his early teens. He had been packed off to the Mid Yorkshire Institution for the Mentally Defective (later known as the Whixley Colony). He was an orphan, and a brutal foster mother had mistreated him, so the Mental Deficiency Committee decided he would be better off in an institution.

In that first week at Whixley, David discovered that he couldn't hold a conversation with most of the other patients, but that this 'was no fault of theirs'. The majority were over the age of thirty and had serious intellectual difficulties.

In his autobiography, written forty years later, David recalled that time passed very slowly – at least during the hours when he wasn't scrubbing and polishing the corridors and the toilet block. As a 'high-grade' defective, David could be trusted to do hard physical labour in the upkeep of the building. Before long he was promoted to 'the sewing room', working five days a week darning and mending the clothing and linen of Whixley. Ultimately, he would run the shoe-mending workshop, helping supervise other inmates in their work. Whixley's other occupational therapy included mat-making, basket-weaving, pottery, crochet, painting and papier mâché. The boys and men spent their pay on special treats, such as sweets, made available for sale on Saturday evenings.

Those who worked in the Whixley laundry worked as much as six days a week for between halfpence and fourpence a week; the pay rate depended entirely on what the superintendent decided you deserved for that particular week. Not only was the pay erratic, it would be stopped for the most trivial reasons, such as talking in the queue on payday, or if you held out your left hand to receive the pay-packet instead of your right. David thought this was a particularly spiteful act because some of the patients had difficulty remembering left from right.

Each Sunday the inmates were walked *en masse* for five miles through three villages. David felt that it was cruel to expose them to public view in this way and claimed the villagers ridiculed them.

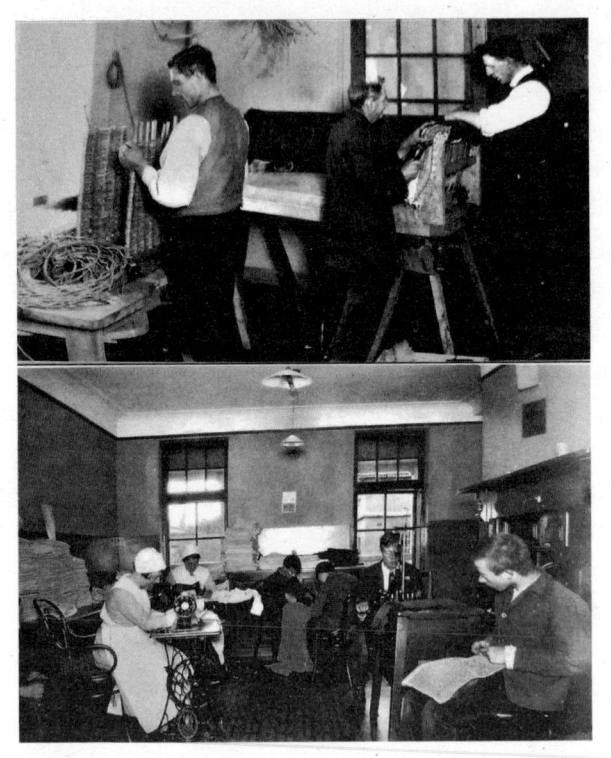

The Occupational Therapy Room and the Sewing Room at the Mid Yorkshire Institution for the Mentally Defective (later known as the Whixley Colony).

The superintendent imposed arbitrary rules, seemingly whenever he felt like it. Trying to recall what to do and when (and what not to do) was bewildering for David and, like most inmates, he couldn't help but fail from time to time. An edict went round that no socks should be worn in bed, and David's meagre pay from working in one of the Whixley workshops was docked for a week when the superintendent pulled his bedclothes off and discovered the offending garments.

The biggest threat that was held over the Whixley inmates was of being transferred to Rampton – the state institution for violent, mentally disordered patients.

> In fact, I saw many patients sent to Rampton during my stay at the institution, and if they went out like lions, those who did come back were like lambs on their return, and no wonder, with the sort of treatment they dished out there.

But the worst rule of all was that the sexes could not meet. David recalled that inmates who were permitted 'outside' for daytime work opportunities, or who were allowed out into the community on licence, were punished if they formed a romantic attachment with a girl. David thought this led to many of the boys and men being unable to form warm relationships if and when they were released. He attributed the failure of his own marriage (which lasted six months) to this aspect of captivity. He also blamed the persistent sexual harassment he experienced from certain older males, which graduated to sexual assaults by two Whixley employees: 'The things he [a staff member] wanted me to do before handing me the chocolate and the writing paper I would rather leave to the enlightened reader's imagination.' David told another staff member of what had happened and the man was dismissed, but the police were not informed for fear of damaging Whixley's reputation. 'Having homosexuality forced on us, as I did, has not helped me any in my later life', wrote David.

When he wasn't at work supervising the shoe-mending, David spent a lot of time helping a lad called Trevor – tying his tie, doing up his laces, combing his hair and so on: 'These attentions made it more or less a full-time job'. In gratitude, Trevor's mother befriended David, and during one visit, told him that she had put Trevor into Whixley because 'youths used to throw stones at him in the street'. She said she suffered terrible anxiety about what would happen to Trevor after her death. In fact, she did die shortly afterwards, and so Trevor was alone. Writing in 1981, David believed Trevor was still in an institution and that this was probably for the best, as Trevor would not have been able to cope alone 'outside'.

***

David Barron's memories, published in 1981 as *A Price to be Born*, are one of the earliest, longest and most detailed works about life in a mental deficiency colony. Various oral history projects undertaken in the 1980s and 1990s also captured eyewitness testimony from elderly people who won their release after the 1959 and 1983 landmark Mental Health Acts. These accounts make a brilliant contrast with the official, legalistic concept of the colony – and what such an institution attempted to achieve.

Mental deficiency colonies were supposed to be unlike traditional lunatic asylums – that's what the authors of the new legislation had claimed. The Secretary of State himself could order transfers between colonies and asylums, if, for example, mental illness had been discovered to be mental disability instead; or if a defective became psychiatrically unwell during their stay at the colony. But for the most part, colony and asylum were intended to comprise two different systems.

The colonies were for the unimprovable and ineducable; by contrast, the 'improvable' (the 'merely dull and backward') ought to have entered the system of 'special schools' within the community. Nevertheless, the colonies, right from the start, did include occupational training and workshops for the so-called 'high-grade defectives', as well as providing shelter and care for the 'low-grade', who could do little or nothing for themselves unassisted.

Keeping males and females apart was a given, and further segregation within the colony ensured high-, medium- and low-grade patients slept in different quarters. In addition, each colony was designed to ensure that visitors were not exposed to the sight of the low-grades, the appearance of whom could cause distress, and might even prompt a visitor to wonder why their relative was being kept in such company. 'Care should be taken to place the buildings for the lowest-grade patients so that they will not be brought to the notice of the parents or relatives of other patients, or to callers', the Board of Control advised. 'The more attractive the mental deficiency colony looks, the greater the parental approval and co-operation.' In a parliamentary debate, Mr Brooks Crompton Wood, Conservative MP for Bridgwater, said that 'the future of the race and their own happiness demanded that they should enter colonies which he [Wood] would testify were a cross between public schools and guild training colleges'.

Even more segregation should divide up age cohorts and separate out the 'wet' and 'dirty' incontinence cases, advised the Board of Control, as well as those with 'dangerous or objectionable habits . . . The greater the sub-division, the better for the happiness of the patients.' And the bigger the colony the better, the Board of Control said, recommending that each should accommodate between 1,000 and 2,000 inmates, because classification and physical segregation was more difficult with smaller populations of defectives. This went against the belief that smaller institutions (of any kind) promoted better-targeted care and control. However, a huge mitigating factor was that large institutions provided economies of scale, especially if you could get the high-grades to do a great deal of the day-to-day labour.

Low-grade inmates were ideally to be accommodated in wings of the main building, on either side of the administration block; but the rest of the colony ought to be based upon the 'villa' model, with individual small buildings set within the large grounds. The Board of Control would often refer to colonies as aiming for a 'village' atmosphere – a large residential population but spread out across a wide enclosed area. The Board envisaged that colonies would become partially self-supporting (as many lunatic asylums had been in the nineteenth century), with high-grade inmates doing much of the work in the colony kitchen, laundry, bakehouse, stores, boiler house, workshops, medical block, mortuary and recreation hall. Each colony should be a little world in itself.

The First World War began four months after the Mental Deficiency Act came into operation, and as a consequence less money and energy were expended than had been planned in creating these ambitious new paradises for the nation's undesirables. With regret, the Board of Control had to accept that workhouses (and later, Public Assistance Institutions) would sometimes have to play the role of colony, as a 'certified institution', until more money became available. A collective groan was sounded by the sector in 1922 when the Treasury sent round a note informing the nation's Mental Deficiency Committees that major retrenchment was under way, and that economies must be found in the administration of the 1913 Act. The Treasury announced that the new Board of Control was annually costing twenty-four times the sum spent on its predecessor body, the Commissioners in Lunacy.

During the war, large unused buildings – whether obsolete institutions or vacant country houses – were much more likely to be requisitioned as army hospitals or prisoner-of-war camps than repurposed as mental deficiency colonies. Prudhoe Hall, ten miles from Newcastle, bucked that trend; the local authority bought it for £13,000 for use as a colony for 348 males and 227 females. Built in 1868 by colliery owner Matthew Liddell, Prudhoe Hall was a neo-Gothic mansion with its own Roman Catholic chapel, set in 300 acres of picturesque land, featuring woods, a glen, a stream and several cottages. It would later become Prudhoe Mental Hospital and continued operating in the mental health sector into the start of the twenty-first century. Many of these former mansions in lovely settings would later sprout additional facilities, including schools, hostels and accommodation blocks, spread out across several sites.

In 1917, the early-Georgian mansion Sandhill Park, near Bishops Lydeard, was bequeathed to Somerset County Council specifically so that it could be used as a mental deficiency farm colony. Sitting in nineteen acres, the property had 130 adjoining acres of arable land, timber woodland and farm buildings, stables, a vinery, greenhouses and its own spring. The building had been used in the early part of the war as a prisoner-of-war camp for German officers. Initially, Sandhill Park was to house high-grade girls and women, plus thirty mentally deficient boys under the age of ten. The Somerset Mental Deficiency Committee advised that Sandhill 'is not intended for young women of immoral tendencies, or for low-grade adult defectives. The Committee are quite satisfied that they can be properly looked after in special wards of the Poor Law institutions.'

Top: Brockhall Colony, Lancashire. Bottom: sketch
of Prudhoe Colony, Northumberland.

Later on, in the 1930s, money began to flow a little more easily into both the mental illness and mental deficiency systems, and a larger number of purpose-built colonies were designed and opened. Nevertheless, shortage of bed-space plus the inability to recruit enough good-quality specialist staff meant that the mental deficiency colony system never performed as the framers of the 1913 legislation had intended.

The Board of Control recommended sites that were rural, yet well-connected enough to towns that staff would not feel isolated. Buildings would ideally have south- or south-east-facing outlooks, and preferably on an incline – a continuation of the Victorian asylum-builders' aim for therapeutic views of nature, sunshine and fresh air. No public footpath should cross a colony: just as Victorian lunatics were kept 'round the bend' – that is, with the main building out of sight at the end of a long curving driveway – so colony inmates should not be visible.

***

Echoes of many of David Barron's experiences at the Whixley Colony are found in the eyewitness accounts of life at Meanwood Park Colony near Leeds, given to researchers Maggie Potts and Rebecca Fido in the early 1990s. Potts and Fido interviewed seventeen elderly men and women released after decades in custodial care at Meanwood following certification under the Mental Deficiency Act. One interviewee had been detained for sixty-three years; and the shortest stay was twenty-five years.

The fearsome sex segregation of colony life also dominated many of their memories, with 'Ernest' recalling:

> We weren't able to mix with any of the female patients, even if we had anyone amongst them. One or two had girlfriends but weren't able to spend time with them . . . We were never told the reason for it. I didn't see any reason for it at all . . . If you have any communication with the girls, it resulted in the person going before one of the senior physicians and losing our privileges and money which we got on a Friday.

During recreation hour 'Grace' recalled that 'all the girls used to have to sit on one side. You couldn't go near the boys. They were watching you . . . I wasn't bothered 'cos I didn't want men 'cos I'm not very fond, but I felt sorry for other girls.' Leisure pursuits at Meanwood included concerts and performances and watching films in the large hall, Highland dancing or dancing to records or the radio, knitting, sewing and playing cards.

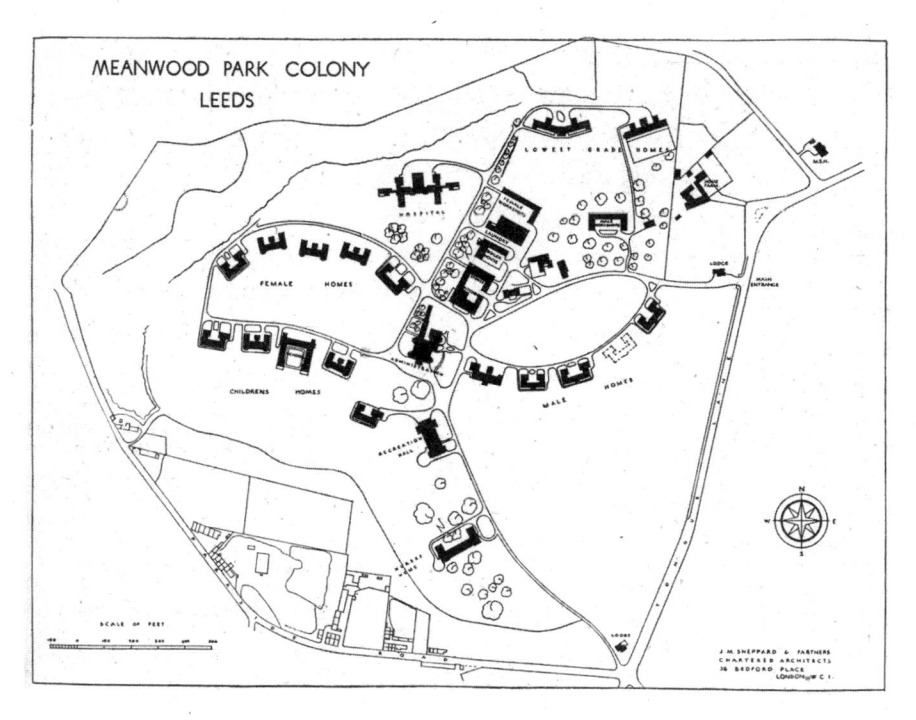

A world unto itself: plan of Meanwood Park Colony in 1941, showing the ideal topography of such an institution.

The interviewees told Potts and Fido that they were under constant surveillance and that no privacy was allowed. There weren't even any locks on the toilet cubicles. Staff intercepted and read ingoing and outgoing correspondence as a matter of course. Items such as towels, combs and sometimes even underwear were communal, and the theft of any personally owned belongings was common and distressing. Ernest said that these conditions caused unhappiness, anger and frustration: 'The villas used to get very noisy. There used to be quite a lot of quarrelling and fighting.'

Punishment for meeting a member of the opposite sex was to be locked in a side room in Villa 8 for girls, or Villa 17 for boys, and later having to scrub the floor wearing only underwear. According to 'Sally',

Villa 8, it were a terrible villa, with all the violence . . . One big stout girl, she'd come out of Rampton. She used to go for all the staff. That's why they were all frightened of her.

However, 'Frank' said that it was pretty easy to abscond from Meanwood – you simply wandered off and slept rough until the police found you. Grace recalled one girl who would go up her villa's chimney and turn up at the local town hall to hang around, all blackened by soot.

At Meanwood, staff segregated the 'low-grades' from the other youngsters; a sense of shame attached itself to them, the elderly interviewees recalled. There was no sex segregation of these most severe cases since it was assumed that their physical disabilities ruled them out of seeking sexual or romantic liaisons. As at Whixley, the capable inmates worked hard in caring for the least capable. Potts and Fido surmise that one of the few outlets for genuine affection and attachment within the colony arose in these relationships between carer and cared-for. Girls were not paid for this type of physically strenuous and exhausting labour, and only in later years did Meanwood bring in equal pay for boys' and girls' chores within the colony.

Another inequality that Potts and Fido infer was that while the boys enjoyed active and organised games, sports and seaside trips, girls experienced less varied and more sedate leisure activities. The authors wonder whether this may have been because of a perceived greater need to tap male energy and 'inappropriate' feelings by keeping them physically active and engaged.

The Meanwood uniform for girls comprised a frumpy skirt, white apron and cap, and heavy boots; at weekends an old green dress was permitted. 'We looked awful', the female interviewees agreed. They believed that this was done deliberately in order to make the girls look unattractive, though the Meanwood boys also complained of the shabbiness and drabness of their own institutional clothes. For its part, the Board of Control – *à la* Mr Brocklehurst – advised colony management to 'ensure clothing is not of unduly expensive quality and not discarded until past repair'; some girls even had their long hair cut off upon admission – someone had been reading their *Jane Eyre*.

Other parsimonies involved no hot meals being served in the evening, and food often going 'off'; baths were weekly, with a suicide watch mounted for each inmate – an aside that illuminates how colony life could impact on the lonely and the desperate. Visiting days and times were inflexible for relations, and totalled just two hours once a month.

Swearing and cheekiness brought about punishment; so did incontinence, even when this was the result of a physical disability. And if a meal wasn't fully eaten up, the leftovers would be served up to the refuser the next day.

As David Barron noted at Whixley, many staff at Meanwood were humane and approachable – others brutal. 'Joe' recalled colony staff, doctors and nurses knocking children about, and claimed that by speaking out, he had managed to get one male nurse dismissed. Sally spoke of being straitjacketed and on a separate occasion being placed in 'the dark room' – a notorious Victorian lunatic asylum practice. Frank said he had been punished with an injection but did not name the substance. Bromide and paraldehyde were legal to use, as was a substance called Croton Oil – a laxative that caused powerful stomach cramps and which was given at Meanwood as a punishment, quite legally.

At other colonies, punishments could include a fortnight in bed, in isolation, with no occupation or reading matter; being issued with sackcloth or canvas nightclothes and bedding; and for 'high-grade patients', being sent for months to the refractory ward in the company of 'low-grade' inmates.

When the Board of Control swept in for their annual check-up on conditions, these visits were by appointment, and so, the interviewees allege, Meanwood was made fully shipshape in good time – an accusation that Victorian lunacy-law campaigners had also made against asylum-keepers when an inspection was pending. Sally and Grace both said that the Board therefore never got to see that inmates often lacked basic items, including toilet rolls, soap and towels. 'Deceitful' is the word Sally used for the process.

Peggy Richards recalled the violence of staff towards youngsters at the Royal Western Counties Institution at Starcross, Devon, in the mid-1940s. 'The nurses beat me and hurt me for nothing. They used to punish us by putting a wet towel around our necks and twisting it until we nearly fainted. I often ran away . . . Often when they brought me back they gave me drugs and beat me. Sometimes it got so bad I had to run away just to prove to myself I was sane.' At fourteen Peggy was moved to Moss Side Hospital (later renamed Ashworth) to an even more brutal environment: 'One day there was a girl shouted at a nurse. The nurse thought it was me. She jumped at me, ripped the front of my blouse off, and got me on the floor. She put her knee on my throat and nearly throttled me.'

Mrs Leah L'Estrange Malone, chair of the London County Council General Purposes Committee, wrote to *The Times* to express her upset when she visited a mental deficiency colony, which she did not name. Her language is distasteful to us, but her shock seems genuine:

> I talked to an enchanting little boy of about five, obviously the pet of the male nurses, who had been certified at the age of about three. He was running about among the adults who, as a result of their affliction, were often very grim, sickly and unsightly in appearance and movement. Twice a day he saw the pathetic little band of dwarfed, twisted, hideous little people, both adult and children, being led by the hand by a couple of frightened-looking young probationer nurses for their daily exercise around the grounds. From the background, from a segregated villa, came the screams of the most afflicted, who can never leave their beds. Ought children such as these to be certified almost in babyhood? How many children and young people of this kind are there in this country leading this death in life?

Mrs Malone was a woman with some power – and at the very least, a voice. But her detailed narrative of the horror she witnessed has a curiously underpowered tone in this letter. Did she feel defeated by the enormity of the problem she had uncovered?

For Peter Whitehead, his first nights at St Joseph's School, Sambourne, Warwickshire, meant shivering under the bedclothes in the 'open-air dormitory' there. Taking the notion of fresh air way too far, the school's dormitory was a veranda, with one side largely open to the elements. St Joseph's was the junior facility of Besford Court Catholic Mental Welfare

> **BESFORD COURT CATHOLIC MENTAL WELFARE HOSPITAL**
> CERTIFIED BY THE BOARD OF EDUCATION, THE BOARD OF CONTROL, AND HOME OFFICE
>
> Telephone: PERSHORE 74 & 75
> Telegrams: "PRACTICAL PERSHORE."
>
> Stations: DEFFORD L.M. & S.R. 2 MILES
> PERSHORE G.W.R. 4½ MILES
>
> Resident Manager: The Right Rev. Monsignor Thos. A. Newsome
>
> Secretary,
> Board of Control,
> Caxton House West,
> Tothill Street,
> Westminster. S.W.
>
> BESFORD COURT, Worcestershire.
> 7/12/31.
>
> B. C.
> 9-DEC 1931

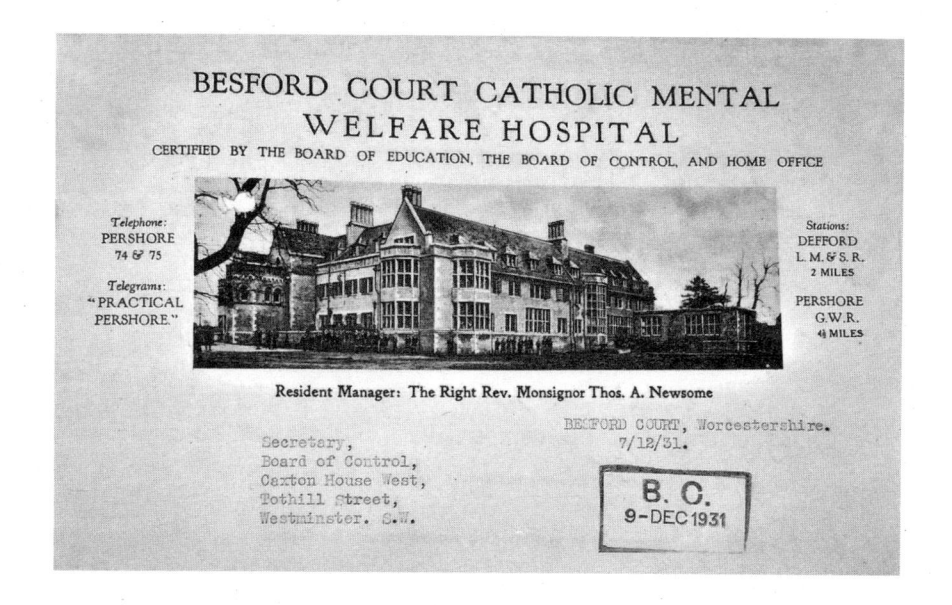

Hospital, twenty miles away, in Worcestershire, and Peter had arrived there in the January of 1937.

Reveille was 6 a.m., then a wash in cold water. Mistakes at Catechism led to a caning. When he transferred across to Besford Court, for older boys and men, the leather strap was frequently in use as punishment. If anyone was caught talking in the dorm after lights-out at 9 p.m., all inmates were hauled out into the cold to do exercises or to stand still for up to two hours in the cold night air. Other punishments included having to stand still on a low wall in the exercise ground for up to three hours; and having to kneel on a pile of marbles with bare knees. Those who absconded and were brought back had the humiliation and physical discomfort of having to wear only a towel around the midriff, for several days.

The baroque sadism of such penalties would years later be documented in television and newspaper reports. In 2004, three men revealed to the *Newcastle Evening Chronicle* the violent punishments they had received from staff at both Besford and St Joseph's School in the 1940s and 1950s; they were among forty young males (one as young as seven) sent to the facilities by Sunderland local authorities. The boys had had no idea that they had been classified as mental defectives – most were from chronically impoverished backgrounds, some had committed petty crimes, some

Psychological tests being administered to boys at Besford Court.

were orphans, some were persistent truants. One told the *Chronicle*: 'I went to the grave of the teacher who made me kneel all night. I looked at it and spat at it. I felt extreme hate for him.'

***

As mentioned earlier in this chapter, it was hoped in Whitehall that mental deficiency colonies would become to some degree self-supporting, since much of the labour power of running them would be free. John Platts-Mills was the QC who would undertake the greatest amount of mental deficiency legal casework, and in his 2002 autobiography *Muck, Silk and Socialism: Recollections of a Left-Wing Queen's Counsel* he wrote that he was in no doubt that the bulk of the certifications he helped to overturn had come about in order to supply the colonies with cut-price labour.

> The institutions procured magistrates to certify sufficient females as mentally defective to satisfy their needs for domestic servants. One popular method was certifying as 'morally defective' women who gave birth to illegitimate children. They could be indefinitely detained and put to work

for as little as two shillings a week. Another tactic was to take a child who had grown up in care and state that she was not mentally fit to live in society.

That's an extraordinary accusation – that sections of the magistracy were corruptly supplying colonies with carelessly, or maliciously, miscertified girls and young women.

It is difficult to know the extent to which Platts-Mills was being facetious when he wrote that one of the recommendations in the 1929 'Wood Report' of the Committee on Mental Deficiency actually urged colonies to run along these lines. He was referring to this paragraph:

> An institution which takes all types [of defective] and ages is economical because the high-grade patients do the work and make everything necessary, not only for themselves but also for the lower grades. In an institution taking only low grades, the whole of the work has to be done by paid staff... The high-grade patients are the skilled workmen of the colony... Laundry, kitchen and housework employ many of the female patients, a few can be used as staff mess-room maids and many are glad to be allowed to help, of course under supervision, in caring for the lower-grade children.

(For what it's worth, by the way, Platts-Mills's view on the origin of the mental deficiency legislation was that the Act of 1913 'gave arbitrary powers of detention' because 'if women defectives were allowed to be free, the next generation would be swamped with subnormal babies. The best disproof of this proposition is that it was not so from the last Ice Age to 1913, nor is it so today. The other disproof is that orphans from any class are not more inclined to produce defective children than are the daughters of established families.')

Whether or not magistrates were 'in' on keeping up a supply of cheap labour to the colonies, there is plenty of evidence that girls, in particular, were worked extremely hard during their detention.

The case of Dora Thorpe led to questions in the House of Commons and tabloid headlines in the mid-1920s, one of which read, 'Tragedy of Tortured Girl: State Slaves Six Shillings a Year. Dora Thorpe Must Be Freed'. Dora had been a persistent truant and during a hearing into her absenteeism she called the School Board inspector a liar. The magistrate sent her, aged twelve, to a residential school; and from there, without her parents' consent, she was passed, at the age of sixteen, straight into Stoke Park Mental Deficiency Colony, near Bristol. She had been declared feeble-minded. At Stoke Park, Dora worked as a weaver at a hand loom from 8.30 a.m. to 4.30 p.m. for an annual salary of six shillings.

> # TRAGEDY OF TORTURED GIRL
> ## STATE SLAVES SIX SHILLINGS A YEAR
> ### Dora Thorpe must be Freed
>
> WHAT, without hesitation, can be described as a tragic case of cruelty to a girl, calls for exposure and comment. The case we refer to is that of Dora Thorpe, who has spent the last fourteen years as an inmate of Stoke Park Industrial School.
>
> It is an amazing and incredible thing that people in these days can be deprived of their liberty and hidden away in a life of torture and slavery. Dora Thorpe has committed no crime, but apparently she has been sentenced for life.
>
> Fourteen years ago, when Dora was attending school at Islington, her mother moved to Wharfedale Road, and sent the little girl to a new school. Shortly after that she received a summons because Dora was no longer attending the Islington school.
>
> **Bumble the b underbuss**
>
> She was brought before the magistrate, and she so far forgot herself as to call the superintendent a liar. Her daughter Dora was then taken away from her, and the superintendent asked for, and obtained, an order that the child should be detained until she was sixteen.
>
> Clearly this was a miscarriage of justice. Let us see what were the dreadful consequences.
>
> She was then thirteen years of age. The order was that she should be
>
> One doctor after another confirms the finding of the first doctor who certified this child. It requires a great deal of courage on the part of a medical man to contradict one of his predecessors.
>
> What sort of a life, then, is this poor girl compelled to live? It would be a mistake to suppose that she is treated as an invalid or a patient, and nursed or coddled.
>
> She is employed at weaving, and she works at the loom from 8.30 in the morning till 4.30 in the evening, with an hour off for dinner. Even this hour does not belong to her, for she has to give up half of it to waiting on the secretary of the institution while that lady has her lunch.
>
> *Dora Thorpe, from a recent photograph*
>
> The State, therefore, employs this girl—whom it has branded as a mental deficient—at hard, constant and strenuous work, which she could not do if she were deficient.
>
> But to be quite fair to the authorities they certainly pay the girl for
>
> It is not too much to say that this is torture and slavery, and a disgrace to the State.
>
> Two years ago we raised the whole question of Dora Thorpe's incarceration. We did not succeed in obtaining her release, but, apparently, the authorities felt perturbed at our criticisms and this year Dora has been given a holiday, and allowed to come home for a few weeks. This is on condition that she returns on the agreed date.
>
> **Home Office challenged**
>
> When we heard that Dora was at home with her mother, a respectable and hard-working woman, who lives at 117, Chalton Street, King's Cross, we decided to raise the issue with the Home Office as bluntly as we could.
>
> We, therefore, invited Dora to go to see Dr. H. G. Faulkner, an eminent specialist of 20, Park Crescent, Portland Place, W. 1. Here is the certificate that he gave:
>
> "I have to-day seen Miss Dora Thorpe, and questioned her, etc. I personally can see nothing at present wrong with this girl. She appears to be bright and intelligent, and I see no reason why she should not be employed at household work."
>
> We say plainly to the Home Secretary that, in view of this certificate, Dora Thorpe cannot continue to be held a hopeless prisoner. We invite Sir William Joynson-Hicks to act

The newspaper agitation worked: Dora, by now twenty-six years old, was re-examined, found to be of perfectly ordinary aptitude and behaviour and was released to her mother, finding a steady job almost immediately.

Elsewhere, 'Ivy' spent five years in a colony washing 'fouled' sheets for one shilling a week. 'Jane' worked in a colony laundry that was so efficiently run it took in a great deal of private work from the local area – making a profit that was not reflected in Jane's pay.

Goods and produce for sale on the open market were a feature of some colonies. At St Catherine's Mental Deficiency Institution, on the outskirts of Doncaster, the male patients drained and fenced the entire estate and had sixty-five acres under cultivation – growing wheat, oats, sugar beet, potatoes and other vegetables. It was so successful an endeavour that it became a registered producer under the governmental Wheat Commission and the Potato Marketing Board.

It is clear that huge numbers of allegedly 'socially inefficient' people were instead proving to be good, productive workers. At no point did this become more apparent – the entire set-up so farcical – than during the Second World War, when call-ups for military service impacted on staffing levels within the institutions. The Board of Control experienced a rising tide of requests from mental deficiency colonies asking permission to place their highest-grade patients into staff positions, with pay that went beyond pocket money (but which came nowhere close to the standard rates for the jobs).

Trades union bodies were not happy with this trend. One, the National Union of County Officers, complained in October 1942 that supervising and teaching an inmate was crucial if they were ever to be returned to society. What's more, the unions believed that staff were being laid off in order that this much cheaper form of staffing could take place. Employment of inmates 'should be educational and not simple drudgery. This educational occupation should be under the supervision and direction of qualified and registered instructors.'

So Henry C— was paid ten shillings a week for a forty-hour week painting, decorating, carpentry and boot-mending at a colony near Potters Bar, Hertfordshire, in 1940 (this was around twenty times less than a 'normal' man would have earned for such work). Violet B— got the job of laundry assistant, 'where she is in her element', according to the financial secretary of the Birmingham institution where Violet was being detained. She was 'a born leader . . . the other inmates obey her – she doesn't want to leave . . . doesn't want to go home.' In Easthampstead, Berkshire, inmate Cyril C— got the role of engineer when the incumbent of that post was called up to fight; the institution paid Cyril a miserly five shillings a week. Most extraordinary of all, in 1942 it came to the Board of Control's attention that apart from the superintendent, the entire staff of one colony were current or former inmates. 'In theory, this should be full of pitfalls; in practice, it seems to work amazingly well', an internal memo records.

(Sadly, and annoyingly, no other paperwork on this unidentified institution and its highly unusual experiment is to be found in the relevant Ministry of Health file at the National Archives.)

By this point, over 46,000 people had been placed in the nation's seventy or so mental deficiency institutions; a further 70,000 were living under guardianship or statutory supervision within the community. So, let's look now at some of the types of people who, after 1914, found themselves being declared mentally defective on grounds that didn't have a great deal to do with their intellectual abilities. To what extent did the colonies and guardianship act as dumping grounds for 'difficult', 'troublesome' or simply 'different' youngsters? The ones we didn't want proliferating.

9

'SHE HAS A BAD NAME IN CLIFTON HAMPDEN': BESSIE B— AND THE POLICING OF FEMALE SEXUALITY

Staff at the Oxford Voluntary Association for the Care of the Mentally Defective were panicking. It was February 1919 and twenty-four-year-old Bessie B— was about to be discharged from Oxford Prison, having served a sentence for vagrancy. Two years earlier, when she was in gaol for child neglect and desertion, doctors had examined her and found her not to be liable for detention as a mental defective: although she was the mother of an illegitimate child, born when she was sixteen, she appeared to be perfectly capable of earning a living, had never taken Poor Relief and had done reasonably well during her school years. However, in the intervening time between prison sentences, she had given birth to two more children, and was now pregnant with a fourth – all by different fathers. She had contracted syphilis, and the police constable who had arrested her for vagrancy late one night in Southampton in April 1918 said that she was clearly destitute and her physical condition was 'very dirty'. This was her second arrest for rough sleeping. It turned out that the PC had in fact discovered her in an army lorry in the company of soldiers. The local workhouse master claimed: 'A Scottish soldier who she accused of being the father of her last child stated that he knew of at least thirty men in his regiment who had had immoral relations with her. No affiliation proceedings were taken against the alleged father of her last child.'

The Oxford Voluntary Association was determined that such a creature should no longer be permitted to pollute her environment, nor to have any more children in whom she took no interest. (Her fourth child would be born with syphilitic ophthalmia.) They wanted her declared mentally defective, placed in Abingdon Workhouse as a 'place of safety', and detained there under certification. Here, she would be able to continue with the injections to clear up her venereal disease.

The Board of Control duly sent down a medical specialist who tested Bessie and found that her arithmetic was fairly good but that her reading was 'stumbling' and filled with guesswork; he estimated it was probably at the level of a child of around nine or ten years of age. Bessie had failed to make up a sentence using three simple words given to her, or to be able 'to define common objects of everyday life'. He noted

> the half-heartedness of her interest and effort . . . Asked about her previous unfortunate career, she said she could not restrain herself on the occasions on which she fell into loose ways and attributes her periods of abandon to her discovery of deception on the part of bad mistresses who in promising to help her intended to place her into institutions.

This last line suggests that while working as a servant, Bessie suspected that middle-class ladies might target her as a suitable case for permanent detention. Middle-class ladies' aversion to Bessie B— and her kind is amplified under the heading on her certificate 'Facts communicated by others', in which we read that

> Miss Beatrice Hatch of Christ Church Oxford tells us she has seen her behaving badly and immodestly in the streets as if she had lost her head. That she could be quite rational and have good manners until she was crossed or thwarted and then she loses her temper. That she is absolutely callous with regard to her children and absolutely careless of her immorality. That she lacks the power of consistent application.

(Miss Beatrice Hatch had been one of Lewis Carroll's 'child-friends', and was one of the infant girls he had photographed nude. She wrote in his obituary in 1898 for *Strand* magazine that 'girls little and big were admitted into friendship at once. Sometimes on the seashore, sometimes in a railway carriage, the magnetic power began, and sometimes, in many cases, continued for life.' Little Miss Hatch had now grown into a bastion of Oxford society chasing down the unladylike public behaviour of those who'd enjoyed less gilded girlhoods.)

Bessie's mental deficiency certificate, signed in the autumn of 1919, additionally states:

> She is untruthful. She cannot read properly. She does not know how many half-crowns in a pound. She sometimes thinks of her children but has no real maternal affection. She admits she has been cruel to them. She lacks power of continuous application. She is of quick and violent temper on occasion. She refuses to work.

Another statement claimed she was 'crafty in her answers . . . She shows no regret for her past life, nor any desire to be different.'

The workhouse master had to concede to the Board of Control that the workhouse medical officer could not report that Bessie was of violent or dangerous propensities, but he did agree that she was 'a fit and proper person to be detained in an institution in her own interest'. To be fair to the Board of Control, they tossed all this documentation back to the Oxfordshire Mental Deficiency Committee with red-pen comments where the paperwork was questionable: 'Particulars of mental symptoms are requested', the Board wrote, aware that Bessie's certification as 'a moral imbecile' probably ought to rest on more than just 'bad character', or 'bad temper'. Or poverty.

As noted in chapter 7, the ambiguity of the moral imbecile sub-category was causing problems: was Bessie of poor enough intellect to be unable to look after herself or earn a living? Or was her inability (or unwillingness) to be monogamous and a loving mother evidence enough of imbecility? If she was about to lose her freedom, to be detained indefinitely, the distinction really did matter.

Another grey area in the drafting of the legislation was an unmarried pregnant woman's receipt of Poor Relief. Bessie had 'got away with it' for seven years because she had demonstrated that she was able to earn money – working as a domestic servant. She had not taken any relief payments from the parish. But the minute the Poor Law authorities placed her in Abingdon Workhouse (an institution the locals called 'the Grubber') she was automatically in receipt of parish relief. This conundrum (we will force you into the category of recipient of Poor Relief by insisting you go into the workhouse against your will and then label you feeble-minded as a result of this) would flare up into national concern in a landmark case of 1955 (discussed later).

Bessie was statutorily entitled to a review of her case after one year of detention. The report written on her first anniversary (in September 1920) states that she remained 'a moral imbecile' but that 'she has behaved herself well during the past twelve months... I think it advisable that she remains here for a further term'.

Four years later, the new master of the Grubber wrote that he had interviewed her for her five-yearly re-examination, during which she told him she was so unhappy in the workhouse that she planned 'to bolt', and that Oxford Prison had been preferable. 'She stated that it was the fashion for girls to have babies and every girl should have one.' He tested her reading and said she had great difficulty, mistaking 'necessarily' as 'necessity' and 'importance' as 'impertinent', and she could not pronounce the word 'paramount'. It is possible that Bessie had mild dyslexia (not recognised in the 1920s, and possibly the cause of the certification of many youngsters as high-grade mental defectives down the years). The new master continued:

> She has a bad name in her native village (Clifton Hampden) and her parents do not wish her to be released. I am of the opinion that she would become a prostitute if allowed to take her discharge. She attempts to disorganise the discipline of the institution, and for the sake of the other inmates it would be advisable that she should have greater supervision than it is possible for her to get at this institution.

This was a veiled request to the Board of Control to find a colony place for Bessie.

The Board of Control remained unhappy with the reasons for Bessie's detention: yes, Bessie's four illegitimate children made her an undesirable. But on the other hand, corroborating evidence of feeble-mindedness was weak. Hundreds of thousands of British people either could not read at all or could read only with difficulty – they didn't require lifelong detention.

The Board wrote to the Oxford authorities that 'with this document in its present form the Board will be unable to continue the Order of Detention in this case and they will have no option but to authorise the immediate and unqualified discharge of this patient from the Abingdon Institution.'

However, they admitted that this too was not a satisfactory solution: 'The Board think that such a step would probably be attended with disastrous results, especially having regard to the history of this case.'

As we have seen with the case of Dora Thorpe, as early as 1925 the popular press had seized upon the newsworthiness of youngsters unfairly shut away under the 1913 Act. However, Bessie, as a presumed sexual delinquent, would have proved a hard sell to the tabloid-reading public, and no champion rode out for her from Fleet Street.

A compromise offered itself, and for the next twenty years, Bessie would live out on licence in Hove, in Sussex, working as a housemaid for retired nurse Mrs S—, under the watchful eye of the nation's biggest mental deficiency voluntary association, the Brighton Guardianship Society. Bessie lived rent-free, and earned £30 per annum, plus a shilling a week as pocket money. She saved a total of £130 in her Post Office account.

The Mental Deficiency Act had not envisaged much usage of the guardianship or supervision clause; but by the early 1920s, it had become apparent that a lack of places in colonies and the ongoing parsimony from the Treasury meant that associations such as the Brighton Guardianship Society were increasingly trusted to offer a way of operating at least some level of control over certified youngsters, so long as they could source the right kind of families and individuals to shelter them. By 1926, Bessie was one of 618 cases on the Brighton Guardianship Society's books – the organisation taking cases from all over the southern part of Britain.

In April 1927 the Brighton Guardianship Society found that Bessie was doing well and 'appears to be in good care … a good worker'. Less impressed was the medical officer sent out by Oxfordshire's Mental Deficiency Committee, who wrote in January 1928 that Bessie was 'feeble minded, talkative and inclined to giggle at very little'.

This clash of opinions continued for several years of check-up visits. The Board of Control's doctor concluded Bessie had a 'fair' general education and read and wrote well: 'She says her downfall was caused by a desire to wander and to have a good time, which led her to the streets.' He added that she and Mrs S— got along very well, and the latter trusted her entirely with the running of the house, and knowing how to pay tradespeople and bring home the correct change. Bessie spent a great deal of her free time with the Salvation Army, and one of the local female Salvationists would go for long country rambles with Bessie on her day off. 'I found her very neatly and suitably dressed this afternoon when I called ... She is very happy and comfortable with this lady who gives her a very good character.' Bessie was anxious to see her parents at Clifton Hampden, now that her father had apparently forgiven her for her former life. She also wanted news of her children – two of whom had been adopted, with two remaining in the Grubber.

But when the Oxford medical officer visited three months later, he was underwhelmed. When quizzed, Bessie recalled details about many of the various inhabitants of Clifton Hampden but she forgot that the Thames flowed through the village: 'She then blushed and said, "Have not you come to bring me a Christmas box" and giggled and ran from the room.' He wrote on his form that she was 'feeble-minded, answers questions intelligently but grins and moves excitedly while being spoken to. Mrs S— says she is still unduly fond of long conversations with men.'

One of the questions on the Board of Control's inquiry forms was 'Is there risk of procreation or marriage?', and on Bessie's form it was stated:

> Very great if it were not for the careful supervision of the guardian ... Mrs S— gives a very good report of Bessie in every respect except that she is still very obviously attracted by and attractive to men. She has had to warn youths at the Salvation Army of this and exercises great care in regard to Bessie, who does not seem to resent her control ... She finds it necessary to exercise close supervision to avoid her making friends with men.

In 1938 Bessie was in her early forties, and although nobody was as crude as to say so, her fertility was probably at an end. Mrs S— now asked if Bessie could be discharged from her certification, even though Bessie had said she'd like to stay on as her housemaid, as a free woman. Mrs S— had visited Bessie's parents (without Bessie) and told the Board of Control that they were 'nice, respectable people'. Mrs S— wrote that 'there is nothing obviously defective about her. She is now a woman of forty-three, steady, and, I should think, perfectly reliable and trustworthy'. Sadly, Bessie then completed a disastrous Board of Control test in December 1938, getting her own date of birth wrong by eleven years and exhibiting a poor short-term memory. 'Her mental age works out at almost exactly nine. She is a pleasant-mannered, nicely behaved woman, obviously rather excitable and apt to give up answering any questions which presented any difficulty', the Board wrote.

The Brighton Guardianship Society also noted some deterioration since her previous testing, stating that

> her excitable manner during the examination points to some mental instability ... She is ... considerably lacking in ordinary initiative and judgment and of understanding in things which are above her sense level ... She would not be able to resist the influence of anyone who wished to exploit her. I regard her as still feeble minded within the meaning of the MD Act and with her facile temperament, her instability and want of initiative, I feel sure that if she were discharged she would not be able to take proper care of herself or her affairs if she were to get into designing hands.

Nevertheless, one year later, in October 1939, Bessie was discharged from certification and officially declared 'normal'. I am delighted to reveal that she married in 1943 and died aged eighty-eight in Sussex in 1982. I hope she managed much giggling across her forty-plus years of freedom.

Bessie's can be read as a moving story of reform and rehabilitation – finding peace and a position in life in place of chaos and misuse by men. Or we could read it as a tale of an individual being 'broken in' – like a Victorian penitent – tamed and made useful for a society that would not tolerate her wildness. She had been no more promiscuous than the men who had impregnated her (the fathers of her children were never chased up to pay maintenance for their care), or the lorryload of soldiers who had allegedly made use of her. The patchwork of 'evidence' of her mental capabilities is highly contradictory.

It seems perverse to say so, but six-and-a-half years' forcible detention in the workhouse under the Mental Deficiency Act followed by twenty years with an agreeable and supportive employer means that Bessie was one of the luckier women to have been deemed a moral imbecile in these years. Many British families have stories of female relatives who were 'put away' for having a baby out of wedlock.* And when the great mental hospital decarcerations got under way between the 1960s and the 1990s, one of the most newsworthy phenomena was the discovery of elderly women who had originally entered the mental health system under the Mental Deficiency Act because they were unmarried mothers; and who, with no one to champion them, had drifted from a mental deficiency colony into the mental hospital system proper. Perhaps they had indeed developed a mental illness – and no wonder: arbitrarily locked away, their child taken from them, deceived and abandoned by the child's father, shunned by relatives and former friends and neighbours. Or perhaps they had simply become institutionalised – the impact of having their autonomy removed for decades leading to behaviour and a demeanour that to the casual glance was indistinguishable from mental illness.

Patient confidentiality means that it is not easy to access individual case histories of this kind. As psychiatric hospitals closed in the final thirty-five years of the twentieth century, patients' casework documentation often ended up being tossed into skips alongside the fixtures and fittings ripped out ahead of demolition or redevelopment of the hospital building. Local record office archivists who were alerted quickly enough sometimes managed to salvage the paperwork; but closure notices of between fifty and one hundred years mean that only direct descendants have an opportunity to access these records in county archives ahead of the file opening date. Illegitimacy and accusations of hereditary mental defectiveness are still deemed so potentially sensitive and shameful that we understand fully why archivists treat these files with such a high level of caution. Significantly fewer privacy issues pertain when the alleged 'moral imbecility' took the form of stealing, vandalism or vagrancy.

Caroline Hill has recently published online a collection of vox pops undertaken in 1991 in the final months of Starcross Hospital. Many among the medical and admin staff recalled exactly why certain girls and women had ended up in this 'mental handicap hospital'. One nursing sister recalled

---

* See Appendix 2 for the typical kind of hearsay evidence about such patients – comments (which I have anonymised) made in response to a tweet in January 2022 on the subject of women shut away in the twentieth century for having an illegitimate baby.

hundreds of them – sent into hospital because, perhaps, father had sexually abused them and they had a kid . . . Perhaps Mum was simple and she had a baby and they didn't know where to put them . . . Were they mentally handicapped? No, no, they were environmentally handicapped – by the environment around them.

A GP who regularly paid calls on Starcross patients said,

Sometimes they came in just sort of 'in need of care and attention'. They'd be picked up, sort of living rough in Torquay or somewhere. They were brought in and cleaned up and you'd either find them somewhere to go or the family would take them back . . . I suppose their IQ would be somewhere in the seventies or eighties. They weren't severely handicapped, just couldn't really cope with living on their own . . . Two or three at Starcross seem to have been sent there simply because they had an illegitimate child – mentally, well they were maybe not very bright, I mean they weren't that stupid either.

***

As mentioned, the Brighton Guardianship Society was the country's largest mental deficiency voluntary association. Another sizeable and proactive voluntary association was the London Association for the Care of the Mentally Defective. It was staffed almost entirely by women, and their work became very difficult during the First World War since the volunteers were also involved in war work, whether taking over traditionally male jobs or helping out in munitions production. These volunteer women were still interacting with such thoroughly Victorian bodies as the Metropolitan Association for Befriending Young Servants and the Southwark Diocesan Association that undertook 'rescue' work and other 'moral welfare' outreach for young women.

The case committee minutes of the London Association reveal the circumstances of the young people they tried to help through liaison with the London County Council's Mental Deficiency Committee. Sometimes these were requests for institutionalisation by parents and relatives under the Mental Deficiency Act; in other cases, it was assistance – either some kind of employment, or financial or supervisory help – for a youngster. In the fifty-seven-page ledger for 1916, for example, the most commonly cited social phenomena deemed problematic include being unable to work without supervision, being dull-minded and thus unable to earn a living, having 'unsatisfactory' (i.e. impoverished) home circumstances, being epileptic, being a 'fallen' woman, showing 'out of control' behaviour, and vagrancy. The people listed in this ledger were aged between five and forty-seven, and most were female, by a ratio of 8:1.

These London records contain many examples of girls and women who had become pregnant while unmarried, such as May R—, nineteen, who had been in domestic service in Crouch Hill and subsequently gave birth to an illegitimate child in Marylebone Workhouse; she was found another place as a servant 'but has lately given cause for anxiety . . . [and] is supervised by Miss Trimmer's Mission for Women'.

Sometimes simply being stroppy brought a girl to the notice of the London Association – Catherine C—, nineteen, was in the workhouse at Kensington, and had already failed in her domestic service work placement: 'is moody and difficult in temper – has no home'.

Eliza H—, thirty-two, was deemed to be failing on several fronts: she had been notified to the LCC Mental Deficiency Committee by the medical officer of the Paddington Workhouse. Dividing her time between the workhouse and her parents' overcrowded home in Queen's Park, Eliza was noted by the workhouse medical officer as having had

> three illegitimate children. Is affectionate, cheerful, not spiteful, but easily led. She can wash, dress and feed herself, and is clean in her habits. She cannot be trusted in sexual matters and relates her conduct without sense of shame to the other inmates of the [workhouse] ward. Although good at mechanical housework, she is untruthful and requires supervision, refusing to work unless feeling so inclined. It is stated that she has never been allowed to run in the streets. Is unable to protect herself and would run away with men. The parents are not averse to her going away and are willing to contribute towards her support. For her own protection, this is a case for institutional care.

The clincher was that it was rumoured that insanity was present on her paternal bloodline.

The county of Somerset was another area in which energy and resources were put into seeking out mental deficiency cases. (The Board of Control did actually keep a league table of the effectiveness of the nation's Mental Deficiency Committees and voluntary associations, and became exasperated that there was so much regional variation. Somerset, London and Birmingham were at the top of this table of diligence.) The Somerset Association for the Care of the Mentally Defective, along with the chair of the Somerset Quarter Sessions, were horrified at the Treasury's plans to cut back funding for mental deficiency, and in January 1922 sent the Chancellor of the Exchequer a catalogue of recent troublesome youngsters who, they believed, urgently needed placement in a mental deficiency institution; otherwise

> they would drift into the hands of the police, the poor law, refuges, rescue homes, penitentiaries, maternity homes, other voluntary agencies. It is impossible for defectives to be dealt with by these bodies effectively and economically.

High in visibility on the long Somerset list were girls deemed either a moral danger to males, or in moral danger from males:

> A.B., certified feeble minded. Sent to a certified institution in October 1917. Had had a number of situations, neglectful of personal habits, and beyond parents' control. Had been about with men and was in moral danger. Sister had had an illegitimate child. She has improved greatly, can now do housework and other manual work well, but would be unable to stand alone, and needs constant encouragement and tact in managing her.

> M.N., age twenty-five. Has been in three rescue homes and has been treated for VD. Placed in carefully selected situations, but discharged as she was associating with men of low character and had fallen twice in less than eighteen months. She is a source of infection. She is certified feeble minded. Cannot be kept in a rescue home where she now is. Parents in poor circumstances.

A letter from Somerset County Council's Mental Deficiency Committee, dated 29 August 1922, revealed that at one of the county's mental deficiency institutions, at Shepton Mallet, there were three men and twenty-five women: 'The female defectives sent to this institution consist almost entirely of young women who have given birth to illegitimate children.'

The West Lancashire Association for the Care of the Mentally Defective went so far as to employ volunteer females in what they described as 'Women Patrols' – presumably mobile units seeking out suspected feeble-minded youngsters across the North-West.

> E.M., Liverpool, age fourteen. This girl was notified . . . by the Liverpool Education Committee in June 1921 and is a very low-grade defective. The home is a very poor one, and the mother appears to be quite unable to control the defective. She is extraordinarily rough and unmanageable, and her habit of stopping any man she sees in the street and asking him for pennies, conduces to bring her into grave moral danger. She is also a vagrant, and frequently wanders away from home and cannot be found. The father is a sailor and is away from home the greater part of the year and Mrs M is extremely anxious that the girl, who is morally somewhat perverted, should be got into the shelter of an institution before she gets into serious trouble.

We can only surmise the extent to which seemingly precocious sexuality may have been the result of childhood sexual abuse in the cases mentioned in this chapter. 'G.H.', aged twenty, was, unusually, singled out by the West Lancashire voluntary association as having been sexually assaulted, though the volunteers failed to stretch so far as to make an explicit link between that experience and her subsequent 'indecent' speech and behaviour:

> This girl has been in touch with this Assoc. since the middle of 1917. She is an orphan. Her father died of consumption; her mother, who was deaf and dumb, is also dead, and an elder sister is in an institution for MDs. G. H. attended school but was found to be very dull and backward. She was also sullen and untruthful. When her mother died, she went as a servant to a lodging house, where she was the victim of an attempted assault by an elderly lodger. She was after this in a Salvation Army Home and the superintendent there found her too feeble-minded to be placed in service.

She remained in this home for a year and at the end of that time was considered to be in urgent need of institutional care. Her conversation and language were very bad. She shouted indecent remarks to men and boys in the streets.

Many of these appear to be clear (to us, in the twenty-first century) cases where social, and sexual, control was being exerted through the Mental Deficiency Act, the label 'feeble-minded' operating as a catch-all to remove from circulation girls and women whose sexuality manifested itself in just the same way that a male's might. Josiah Wedgwood had argued that the legislation would be used to control working-class girls and women, and he was correct.

Wedgwood had also said that the Act would operate as a blunt instrument against prostitutes. An intriguing, anonymous eight-page document in the London Metropolitan Archives on the topic of venereal disease and mental deficiency incidentally discussed the certification of prostitutes as mentally defective, and seems to bear out Wedgwood's prediction. The paper is undated and may have been some sort of internal memo. It appears to have been written after 1922 and probably before 1930; and it contemplated possible connections between venereal disease and congenital feeble-mindedness. Its author states that women working in prostitution and coming up before the magistrate were a cohort who frequently found themselves assessed under the moral imbecile sub-category of the Act. But all too often, after they had been certified as defective, it would come to light that a significant number of them, once living inside an institution, were found clearly not to be so. For this reason, they were

hard to detain ... In cases of girls of the prostitute class, it might well happen that at first they would obtain some other livelihood but that exposed to temptations they might in time relapse. The really difficult cases are those on the mental borderline who are from all standpoints the most dangerous to the community. If evidence were available that a girl had returned to an immoral life, she could be re-ascertained, but in instances where such a girl has changed her residence to the area of another authority, it is possible she might escape notice for a considerable period.

This type of insider memorandum gives heft to Wedgwood's belief that the Mental Deficiency Act was deliberately framed to be part of a social control model to target prostitutes – the young women who worked in that profession failing to conform to traditional ideas about feminine behaviour, plus showing high levels of recidivism. The Act can perhaps also be viewed as a twentieth-century covert version of the Contagious Diseases Acts of the 1860s, so successfully challenged and thwarted in the 1870s and 80s. The Mental Deficiency Act was now arbitrarily sweeping up vast numbers of girls and women working in prostitution and, following a court hearing, confining them in institutions on highly dubious evidence of defectiveness.

The fact that a female repeatedly returned to this way of making a living fitted well with the idea that recidivism was a sure symptom of feeble-mindedness. Mental deficiency had, after all, been cited by prison medical officers as linked to individuals upon whom repeat punishment had no deterrent effect. At this particular time, it occurred to few people that becoming involved in the vice trade may have been in part the result of catastrophic early-years abuse; or the economic fact of having significantly fewer other options for making a living. In his 1940 memoir, Josiah Wedgwood made this strange throwaway remark, which the London Metropolitan Archives document may shed light on. Wedgwood wrote that

> the Bill had some merits, but it was one whereby prostitutes could be sent to feeble-minded homes to save mankind from infection. I did not know anything about prostitutes . . . but I did know that this was a clear case of Expediency v. Justice. In 1912 I wrecked the Bill on Grand Committee and in the House.

Here, he anticipates the use of psychological medicine to tackle a social problem.

Also in the London Metropolitan Archives is a page of tabulated data entitled 'Women and girls known or alleged to be immoral and notified as mentally deficient, up to 1922'. These statistics were obtained from London prisons, police courts, Poor Law authorities, homeless refuges, and 'misc'. Of a total of 300 surveyed women who had been charged with, or found guilty of, solicitation, sixty-five had been found to have venereal disease;

264 had been certified as feeble-minded, and nineteen as moral imbeciles; and seventeen were deemed 'normal' after all.

I strongly suspect that this table and the London Metropolitan Archives memo that it accompanies are the work of Cyril Burt – geneticist, hereditarian, controversial proponent of the inheritability of IQ, and from 1913 to 1931 psychologist for the London County Council. In 1915, Burt devised a programme of testing children for 'backwardness', and he would be one of the first experts in Britain to set up a 'child guidance clinic'. He published his pioneering book *The Young Delinquent* in 1925 and it subsequently went through several editions. Later, Burt would be one of the prime movers behind the introduction of the Eleven-Plus school exam, and he would be the first psychologist to be knighted.

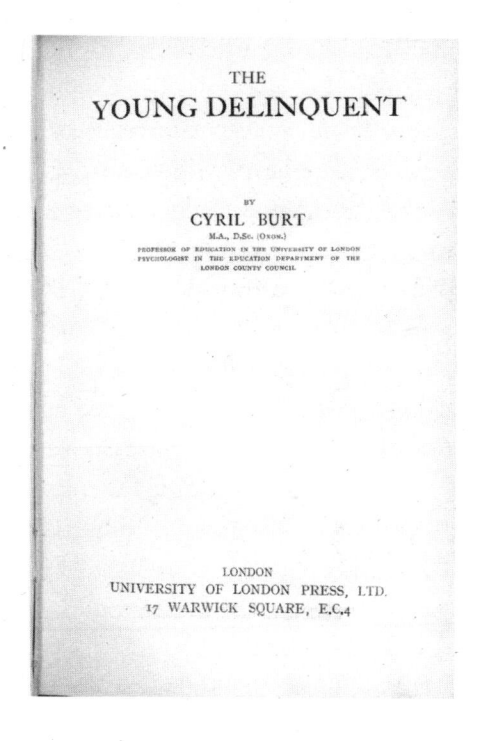

Burt was a crucial figure in the 'New Psychology' – the wave of psychologists who, after the First World War, were part of a cultural shift in education, penal reform and psychiatry. The discipline of psychology was in existence by the 1880s in Britain, but the shocking evidence of the psychical impact of war on troops in the trenches was one driver behind its rise to greater prominence by the early 1920s. But this was a highly uneven takeover from Victorian and Edwardian thinking – and Burt's work on 'sex delinquency' in females manages to contain both modern gestures towards how early-years trauma (not his phrase) impacts on the psyche, alongside thoroughly nineteenth-century judgmentalism and misogynistic double standards.

The year after his influential book, Burt published two articles on his preliminary findings on the psychology of prostitutes. Admirably, he confessed that his sample was small and that in comparison to work undertaken on the Continent and in the US, the material that had so far been gathered in Britain was too 'scanty' to furnish completely convincing conclusions. But he hoped that his findings would prompt more intensive research. The conversations, background research and psychological testing of the London-based girls and women he investigated led Burt to conclude that one overwhelming factor in their turn to prostitution was 'the prevalence of sexual irregularities and illegitimacy on the part of the parents and relatives'. Interestingly, for a member of the Eugenics Society, Burt did not believe 'sex delinquency' was a wholly inherited trait – environment played a large role. He cited the moral and psychological 'atmosphere' in which the girl grew up: 'Bad example, tacit encouragement, and even violation of the girl herself may have been sufficient to initiate her to a career of vice . . . [but] it was difficult to resist the inference that an oversexed constitution was running as an hereditary taint through certain branches of the family.' This is a relatively sophisticated take – for a eugenicist, for the 1920s – and at least raises the issue of childhood sexual abuse. Nevertheless, his research findings on environment were all placed under the sub-heading 'Heredity'. (In chapter 9, the impact of child sex abuse as it is starting to be understood in the 1920s will be explored in greater detail.)

Burt stated that around a quarter of the girls had grown up in 'tenements' that were officially 'overcrowded' (for 1926, that meant more than two adults to a room) and that 'where all ages and both sexes are huddled together within one stifling room, decency is difficult, delicacy impossible, and premature acquaintance with conjugal relations all but unavoidable; an early preoccupation with sexual topics develops very readily, and sexual malpractices are by no means unknown between the girl herself and members of the same household.' And if there were lodgers, 'an easy intimacy with comparative strangers is bound to lessen social reticence and to injure self-respect.' He found that for eighteen per cent of the case histories referred to him 'the first experience of sexual molestation occurred within the child's own home', and in five per cent the assault had been made by her father or brother.

Whether assaulted or not, seven per cent of his case histories had fled home by age eighteen. Around one-third had previously been in domestic service, and Burt was flabbergasted that being trained to be a maidservant was still the main choice of 'reform' that was selected for 'girls in moral peril'.

He did not agree that immorality in itself was a symptom of feeble-mindedness – and found only twelve per cent of the women he was discussing could be liable for detention under the Mental Deficiency Act on moral grounds. But he said he would certify as defective many prostitutes for what he described as their 'temperamental deficiency'. He wrote:

> The most general characteristic of the prostitute is this – she is an emotional and excitable person of the unrepressed or uninhibited type. Sometimes this innate emotional instability is so extreme as to amount to what I have termed 'temperamental deficiency'; the girl needs permanent care for her own protection and for that of others, and should be certified accordingly.

Lawyer and doctor Letitia Fairfield, who knew Burt and his work well, believed in the early 1930s that 'the feeble-minded prostitute is now a thing of the past; the absence of the licensed house of ill fame in England probably accounts for the small number of feeble-minded prostitutes – they cannot maintain themselves in competition with their normal sisters.' Childhood abuse did not cross her mind at all when she wrote in 1931, perplexedly, of a number of cases of girls and women certified after committing crimes, that 'all these girls gave a history of a "difficult childhood", sullen, vindictive, irresponsive to affection, with a marked increase of instability at puberty, showing itself in violence, strong sexual or homosexual trends and incapacity for settled work and a gradual "retreat from reality" into a world of phantasy'. It is somewhat surprising that Fairfield could think of no likely reason for this phenomenon.

Historian Linda Mahood has written about Scotland's revelations and official attitudes towards child sexual abuse and incest after around 1920, when discussion there became more open. She notes that the term 'wandering' in a girl's criminal or welfare notes often indicated a wish to avoid home as a site of assaults. (More girls than boys were described as 'wandering'.) Mahood quotes Mrs James T. Hunter, founding member of the Scottish branch of the National Vigilance Association and a lock hospital director, who predicted that many girls, when assaulted as children, 'grow up to be dissolute women'. Girls would often be sent into some form of custodial control in order to save them from the dangers within their own homes. Mothers would often be blamed for 'moral neglect' – for failing to prevent the assaults.

Cyril Burt's contemporary and fellow pioneer John Flügel was one of the earliest adopters of psychoanalysis in Britain. He, like Burt, had a foray into linking childhood trauma to later behaviour – discussing openly a topic that was still hugely taboo, even in 'serious' publications. Flügel wrote in 1921 that

> although in civilised communities regarded with almost universal condemnation, incest has probably always existed to some extent among certain sections of the population, and the practice of incest among modern white races is undoubtedly much more prevalent than is commonly supposed. A well-known British psychoanalyst assures me that in the exercise of their profession he and his colleagues hear with astonishing frequency of cases of incest, the report of which is otherwise suppressed. Particularly is this so as regards children ... It is startling to note in this connection that, according to the Chicago Vice Commission, of a group of 103 girls examined, no less than fifty-one reported that they had received their first sexual experience at the hands of their father. Even if we allow a liberal margin for incorrect or exaggerated statements ... these figures would seem to afford astonishing evidence as to the prevalence of incest of the father-daughter type in the towns of America. In this country there is reason to believe that similar occurrences are far from being uncommon.

***

In October 1950 the Society of Labour Lawyers wrote to Labour Minister of Health, Aneurin Bevan, in plain language, highlighting the iniquity of applying the 'moral defective' label to girls and women on account of sexual activity. The Labour Lawyers were very unhappy that the mere fact of a female having an illegitimate child should be a diagnostic criterion for feeble-mindedness:

> The birth of the baby very often brings into prominence persons who have never been considered certifiable, nor have come into contact with the authorities, and indeed, in many cases, they, up to the birth of the child, have been in regular employment . . . [And] the existence of this provision makes it possible for girls who are social problems to be certified because, in fact, those who have tried to help them see no other way of dealing with them . . . It remains, therefore, a purely personal judgment of any doctor or psychiatrist.

It was quite unsatisfactory that there was no official right of appeal or right to challenge a doctor's opinion; and the Society alleged that the one-year and then five-yearly re-examinations were 'mere routine matters'. The Act did permit a parent, relative or friend to commission a certified medical practitioner to make a re-assessment, but 'as the people certified are almost invariably poor, ignorant and of the lowest strata of society, none of them have the faintest idea that such a right exists, and even if the alleged defective did know, it is difficult to see how he or she could get such a report without the means to pay for it.'

In the case of unmarried mothers, the Labour Lawyers pointed out the suffering of the baby itself: 'There are many instances of girls at present detained who are devoted to the children from whom they have been removed.'

Among the cases that the Labour Lawyers cited in their memorandum to Bevan was that of a seventeen-year-old who had been certified after having been found neglected, and suffering with a venereal disease. She absconded from the colony and while on the run, met a man, set up home with him and gave birth to a child in June 1949: 'The Health Visitor reported that the home was squalid but that she was a good mother and devoted to the child and could manage with supervision.'

In February 1950 the man with whom she had been living was put on trial for theft at the quarter sessions. The alleged defective wrote to the court pleading for leniency. In sentencing, the judge stated that he was taking into account that the guilty man had obtained 'the love of a good woman, from which it can be seen that this alleged defective could write an extremely effective letter'. But someone in the locality found out about her past and 'she is now back in the institution and they will not release her.'

The second case history they had discovered was a young woman who had been torn apart from her child. She was deaf, which, the lawyers stated, was likely to have affected her educational attainment. Suffering from venereal disease, she gave birth to an illegitimate child and was certified as mentally defective within eleven days.

> She ran away and was picked up by a police officer, who later stated, 'Why is this girl locked up? If she is bats, I am too.' The girl can discuss, and does, the maternity benefit due to her and enquires into the reasons why it is not received. She is also a devoted mother.

Perhaps most distressingly, it was often – perhaps usually – the parents of such a girl or woman who asked their local authority's Mental Deficiency Committee to ascertain her with a view to certification. This may have been done to avoid the shame that illegitimacy might bring to the family; or perhaps to cover up and deflect from incestuous abuse. If the anecdotal evidence supplied in Appendix 2 is typical, blood relatives were significantly more likely to be responsible for the catastrophic incarceration of an unwed mother than the 'Women Patrols', or the local authority 'rat-catcher'.

By contrast, youngsters swept off to a colony for persistent stealing often had their freedom fought for by angry relatives. Significantly less ancestral shame attaches itself to the people put away as 'morally imbecilic thieves'.

10

'A CROOKED MIND': THE ADOLESCENT THIEF

An entertaining young crook, John D—, was a eugenicist's dream: hearing-impaired and mute, his mother had the same disabilities, his father and grandfather had both been detained in the Gateshead lunatic asylum, and a paternal aunt had hanged herself. Nineteen-year-old John was said to be 'incorrigible' – unable to stop appropriating the money and belongings of others, and punishment appeared to have no effect on him. In an interview he said 'a funny state of his consciousness' aroused his desire to repeat these offences and that he could not resist, even though he knew he would be punished. His deafness and inability to speak made it difficult to understand some of the situations he found himself in, he told the authorities. (John didn't lip-read, so questions and answers were written down.)

In 1925 he was found guilty of 'false or fraudulent pretences to obtain money' from the Northumberland & Durham Mission to the Deaf & Dumb and of obtaining two shillings and sixpence fraudulently from one William Taylor in Gateshead. For these offences, and with his bloodline taken into account, John was certified 'a moral imbecile within the definition of the MDA 1913 . . . He expresses no shame in admitting various frauds but rather seems proud of them.' They had to admit he was 'of normal intelligence' but his temperament and behaviour meant that he faced a lifetime of being locked up. The certifying doctor wrote:

> He is facile, irritable, bad tempered, and sucks his teeth, clicks his tongue when he is annoyed. He is a plausible liar and emotional. He

admits with the utmost nonchalance that he has been five times in prison. Although he agrees he is doing wrong and realises the penalties, he says he cannot help so doing, not being able to avoid the temptation. He mostly commits crime to obtain money for cigarettes for which he appears to have an inordinate craving. His reasoning is entirely illogical and contradictory. On being tasked with his misdeeds he blames his mother instead of himself. He . . . appears to have no sense of right proportion in life.

John's mother signed the certificate that committed him to Prudhoe Hall Mental Deficiency Colony from where he absconded four months later. He was on the run for five months until March 1926 when police picked him up for impersonating an official donations collector on behalf of the Royal Deaf & Dumb Institution. He had amassed in the region of £170 before being arrested in London (that's around £8,000 in the 2020s). The newspaper report of his subsequent court hearing revealed that 'he was very alert mentally . . . appeared very agitated and stated, through the medium of an interpreter of hand language, that he was not mentally deficient'.

Prudhoe Hall didn't want him back as 'he is crafty and cunning – upsets the Home with his unseemly behaviour and I feel quite sure he will escape again'. So John was sent to Rampton. Rampton had opened in 1912, originally to be an overspill facility for criminal lunatic patients at Broadmoor, but also taking some of the segregated Parkhurst 'weak-minded' prisoners. The facility had been purpose-built in the Nottinghamshire countryside, and by the time John arrived it was housing the following types of inmate: 'idiots', 'imbeciles' and the 'feeble-minded' certified under the Mental Deficiency Act 'who are in addition delinquents'; patients deemed to be suffering from 'psychic disorders' such as hysteria, anxiety, neurosis, dementia praecox (schizophrenia) and paranoia; those suffering psychosis or neurosis; and 'moral imbeciles', as described in the Mental Deficiency Act.

In practice, Rampton tended to be the destination of people within the mental health and mental deficiency system who were proving hard to control – often the most violent people, upon whom no treatment or punishment had worked. Many, perhaps most, colonies tried to use

Rampton as a dumping ground for those inmates whose behaviour caused problems in the everyday running of their institution. As John, and David Barron (in chapter 8), believed, it was held as an ultimate threat over any colony inmate whom the staff wished to intimidate.

Sending John here was an over-reaction on the part of the authorities. But this is what Rampton said of him:

> He presents himself as a crime for others to solve . . . Associated with his deafness and dumbness is an egocentric attitude . . . He regards the reformatory actions of others as pure hostility and an effort to prevent his progress and self-expression. [He had] a desire to revenge himself on his oppressors . . . I regard him as a moral imbecile of the persecutory type.

On the eve of his twenty-first birthday, John wrote the first of several letters to the Board of Control, requesting a transfer from Rampton. He expresses himself on paper in the manner of someone who has spent a long time with the great novelists of the nineteenth century. If Jane Austen had done time in Rampton, she may have written . . .

> Dear Sir, I have been pondering over the lapse of twelve full months as to how I have made any progress to some extent. It is my decision after careful meditation to write you a letter for the second time . . .
>
> . . . I am venturing to ask you whether you have made any intentions for the present on my case, or even the future . . .
>
> . . . It is quite superfluous to reiterate my wish for a transfer either to Prudhoe or to some Northern mental institutions, but I recollect your kind letter in the month of February last which, though a mere acknowledgement, promised me of an opportunity to have a chance which I never had . . . It would please me if your next answer will not contain any design of procrastination . . .
>
> . . . yet I beg to inform you that I now understand the consequences, and possessed in full of my faculties, and I think that I should tell you also that

> I began to understand what a contemptible creature I have been in spite of my kind missioners . . .
>
> It is the content of my heart to live in Northumberland . . .

And so on, for many beautifully handwritten or neatly typed pages.

The Board's reaction? 'Very plausible but lazy, inattentive and impulsive . . . Unfit for removal.' His correspondence was 'the typical moral imbecile communication'.

Nevertheless, the quality of his writing was, privately, noted at the Board, with an internal memo asking, 'He typed this himself?' Rampton replied yes – but 'He has a crooked mind, and release would certainly be followed by a renewal of his sinister activities.'

The years went by. The Rampton Boy Scouts brigade benefited from John's energetic leadership and he became a hard worker in the carpentry workshop. John felt unhappy and vulnerable under the regime, telling the Board of Control, in a passage of astonishing insight into the implications of having a disability in such a hostile environment, that

> it was not very decent that I was obliged to mix and mingle with sodamists [sic], delusionists, moral perverts and the like. Yes, I am still young, but my youth will, I am afraid, die sooner or later . . . Perhaps you know that I am terribly handicapped by being bereft of the power of hearing and speaking. I am almost alone here without a true companion. This accounts also for the difficulties that I meet with when trying to understand not only the beauties of literature, poetry, but also the needs of my fellow man, for they are like a foreign people living in a strange land.

Paraphrasing Selwyn Oxley, a campaigner for the rights of deaf people, whose work he had read about in a magazine, John wrote that

> the deaf and dumb are not bad tempered, not mad, but generally only retarded by late development, and tact is required in dealing with them because they cannot always tell what others say. But a pencil and paper will put this right if only you will see them . . . Deaf men and women have generally made good in those professions which are suitable for them.

Two decades later, Rampton was still the destination of youngsters labelled 'defective' – to the horror of campaigners.* One, David Roxan, would write in 1954:

> The place of Rampton in our mental deficiency system is causing grave concern in the minds of many people. Its patients range from five-year-old children to brutal murderers . . . Half the inmates at Rampton have been in trouble with the law and half have not . . . Every patient is supposed to be violent or potentially violent, yet in many cases, knowledge of the men and women shut inside Rampton makes one wonder how they could ever have been sent there. Often their parents and relatives cannot find out . . . Arbitrary decisions are taken by unknown officials protected by complete secrecy, safe in the knowledge that their actions will rarely be questioned. This is precisely how a police state would operate in this country . . . Rampton is a prison; perhaps differently organised, but still a prison.

In 1934 the Rampton staff noted John's temperamental improvement and declared him fit for transfer to a mental deficiency colony. He arrived at Dovenby Hall Colony, near Cockermouth in Cumbria – another former country mansion bought and converted for colony use in 1930, and in the mid-1930s housing around 300 inmates. Dovenby ran Scout, Girl Guide, Cub and Brownie packs, offered football and cricket for the males, plus indoor games for all, including billiards, darts and quoits. At weekends films were shown and dances took place 'at which both sexes are present under suitable supervision' (said the Ministry of Health inspectorate). There were also the usual large workshops manufacturing goods at cut-price labour rates – pocket money, really.

John's behaviour and demeanour continued to impress. And so in August 1937, John, now thirty-one, was allowed out on three months' licence to one Reverend Kenworthy living near King's Lynn, Norfolk, a placement so successful that it was extended for another six months, and then another six. The reverend was astonished that John should have been classed as a mental defective of the moral imbecile category. Kenworthy wrote to the Board of Control:

---

* It is still going on – see the afterword.

> As a clergyman, I do not understand what 'moral' deficiency means in a man intellectually normal. It would appear to be something of which almost anyone might be accused... He is a man of outstanding ability... He is very ambitious [and] desires to cast off what he calls 'the status of an inferior person'. He thoroughly believes in his gifts as a craftsman and an artist. He has written poetry and a play. He is a good painter and carpenter.

John had also set up the local Scout troop. The people of the village knew nothing of his moral imbecile status, and, Kenworthy wrote, these 'simple' rural folk had taken a while to accept him – as an outsider, and a deaf-mute outsider at that. But he had proved himself and had worked well for a year for a sign-writing and shopfitting firm.

All that notwithstanding, the Board of Control still didn't like John, and following a long face-to-face interview (by handwritten messages, passed across the table) concluded he was a 'facile, egotistical verbalist... opinionated, argumentative etc', and they recommended at least one more year's certification out on licence.

Kenworthy did have to own up to the Board of Control that John may have fallen in love, thereby breaking the conditions of his licence. He had secretly been paying court to local deaf and mute young woman Alice Chapman and had hopes to marry her. The reverend had been cross when he found out, because of the deception; nevertheless, he asked the Board to lift any potential ban on John marrying – marriage would be an excellent thing for him, the vicar argued. However, hoping in the meantime to cool the relationship off, Kenworthy and the Board of Control agreed that John could be reconciled with his mother and brother, by now living in Kilburn, north-west London; and so he moved to the capital, finding well-paid work as a carpenter and painter for the Olympia exhibitions centre.

Kenworthy had an interesting eugenic take of his own, writing to the Board that John was 'of Ulster Scottish descent, and having read Browne's *History of the Highland Clans* [1838], I can believe that their standards of conduct differ from more settled codes'. That, he thought, explained John's earlier love of absconding and his months of vagabondage before the police arrested him in London.

It all went well in Kilburn, and in June 1939 John was recommended for discharge from certification, but to be kept under friendly observation. He married Alice in the summer of 1940, and if I read the genealogy databases right, he lived into his eighties and left behind a large sum of money.

John comes very close to the phenomenon noted by Dr George Savage of the Royal College of Physicians, who, back in 1905, had told the Radnor Commission that the 'so-called kleptomaniacs' that he had come across in penal settings 'very often ... were brilliant people ... One saw a good many of these cases, but beyond the fact of stealing they showed no intellectual defect.' This made it very difficult to make a professional decision about their detention. No matter how 'intellectually brilliant', he said, 'they steal in the most open and stupid manner.' Savage found such people an enigma – they were 'morally insane in relation to property', but in all other respects, they were youngsters with every hope of doing well in the world. Were they mentally ill, morally imbecilic, or simply criminal? Savage simply didn't know.

Dr Millais Culpin of the London Hospital was an early adopter (in Britain) of the psychoanalytic approach to juvenile crime. Writing in *The Lancet* in 1921, to argue against heredity as a meaningful factor in criminality and anti-social behaviour, he stated his belief that

> just as tuberculosis is now known to be often conveyed from parent to offspring by infection and not by heredity, so the evil home influence inseparable from the presence of neurotic or drunken parents is bound to produce effects upon the mental stability of their children quite apart from, or in addition to, the effects of hereditary predisposition ... A child who repeatedly offends in a particular way, who derives no obvious benefit or pleasure from the offence, and on whom punishment has no deterrent effect, is urged by a motive hidden from the casual observer. The offenders are not carrying out conscious wishes, for they are urged by some impulse which they cannot resist or understand.

Culpin believed that kleptomania was 'a psychoneurotic obsession', and that the theory of the unconscious could be enlightening in such cases. Citing a study undertaken amongst juvenile criminals in Chicago recently, some were found to be suffering from 'mental conflict, generally sexual in origin, as a main cause of the delinquency', and there was 'a pathological impulse to offend . . . In our country we have very little experience of this particular application of psychotherapy'. He hoped that this would change. Doreen S—, put away for life at the age of ten for habitual pilfering, was freed because, as in John's case, upstanding members of the community championed her cause. The daughter of a colliery electrician in Deal, Kent, Doreen seemed to change at the age of seven, according to her mother. No reason was suggested. She began sneaking back into school after hours and stealing toys, handkerchiefs, stationery and on one occasion the school's petty cash box, then throwing the emptied box into the sea. She stole from schoolfriends, family members and strangers.

The paperwork for her admission to the West View Certified Institution in Tenterden, Kent, shows the broad range of medical opinions given on little Doreen. Dr D.W. Kirk decided that 'the child has a perverted moral character and punishment has apparently no effect on her. She requires care, supervision and control for her own protection and of others, especially on account of her propensity for stealing.'

But another medical man who had tested her said that she could

> read, write, do simple arithmetic and knows the value of money. She says she steals things because she wants them but will not [do so] in future because she has come here to learn. She requires care and supervision. Clean in habits and able to look after herself. She is a feeble-minded person. High grade. Is in good health.

Another pen in the margin alongside this diagnosis states that Doreen had

> intelligence below the average for her age. False memory and childish . . . She is simple. She makes nervous twitches of the face when speaking. She cannot tell the time. She recognises a penny, but does not know whose head is on it. She cannot give the name of the king or queen. She reads words but without understanding the sense. She knows the number of pence in a shilling but not in half-a-crown. She appears capable of improvement under special care and teaching . . . She is restless and mischievous and distinctly more backward than her sister aged two years younger.

A comment in red pen, by an official at the Board of Control (who reviewed all admissions paperwork), reads: 'Will the medical certificate please add some evidence of moral defect?' Clearly, as in the case of Bessie B— in the previous chapter, the moral imbecile category caused official unease when the young person in question was not of unequivocally low IQ, nor suffering a disability (like John D—), nor descended from a long line of villains. Investigations of the home circumstances of the S— family concluded that they lived in a 'working-class locality... The parents are poor but respectable people. They desired institutional treatment for their daughter as she developed very bad habits, especially pilfering, and became unmanageable at home.' No eugenic worries there.

An internal memo at the Board dated 10 January 1933 describes Doreen's certificates as 'doubtful'. And on the ground in Kent, the local 'Visitors' (that is, a group of magistrates empowered to check up on institutions and inmates, should they choose to do so) were unhappy with the detention of Doreen, who was now about to turn twelve. They wrote to the Board,

> In this case, the Visitors yesterday certified that they are quite unable to find any evidence whatever of mental deficiency, or of moral defect. This is a girl of quite tender years, who, apparently, on the face of the file, was sent to an institution mainly because she tended to petty pilfering...

Whitehall viewed this as impudence, and scribbled in the margin, 'The Tenterden Visitors seem to be anxious to take over the Board's functions!'

Nevertheless, the intervention worked, and four months later Doreen was transferred to a children's home near Sevenoaks and then released. She remained under formal 'statutory supervision' when her parents moved the family to a new larger house upon her father finding a better job. Whatever psychological storm Doreen had been passing through, she came out the other side. She easily caught up at her new school, and social workers deemed her family life 'satisfactory' during follow-up visits. She married at the age of twenty-seven.

If the Kent Visitors had not flexed their muscles, had not been composed of individuals with a libertarian streak, she might have spent decades in an institution. (In twenty years' time, the Kent Visitors would give their forthright views on the unnecessary detention of alleged defectives to the huge parliamentary inquiry into Britain's mental health system.)

Luck played too large a part in the overturning of the clearly nonsensical mental deficiency incarcerations of Doreen S— and John D—. But Jessie H— had only her frightened, poverty-stricken parents. Sixteen-year-old Jessie was sentenced to three years in borstal in April 1917 for larceny – she had stolen a jacket and had a previous conviction for stealing skirts just two months earlier. The medical officer attached to the Clerkenwell Sessions House in central London decided that she was a moral imbecile, 'quite irresponsible – a wandering, unstable character'. He discovered that she had once cut off her hair from one side of her head only, leaving the other half long. Much more seriously, she attempted suicide by self-strangulation and repeatedly threatened to do it again. 'Her temper is variable and she is a bad influence on other girls.'

So instead of borstal, where she would have served her sentence and then been released, Jessie was sent to the Farmfield State Institution in Horley, Surrey. On the admission certificate she was described as 'feeble minded and of dangerous propensities'. Her parents raised no complaint.

Farmfield was a colony that had originally been the London County Council Farmfield Reformatory for Inebriate Women, opening in 1900 for female alcoholics who had passed through the court system. The estate had 374 acres of land, plus two mansion houses, and a number of additional specialist blocks built for inmates. During the War it had been temporarily commandeered by the War Office for 'mental shock cases' before being handed on to the Board of Control as a mental deficiency colony, with places for ninety females with 'violent propensities'. When Rampton opened, these women were transferred there, and in the mid-1920s Farmfield would next become a colony for high-grade 'intractable' males only.

The Surrey Visitors looked Jessie's case over and were not impressed by her family circumstances: her father was unskilled and barely literate; the family lived in one of the worst slums in Lambeth, South London – in China Walk. Supervision at home would be 'inadequate', they decided. She must stay at Farmfield.

It's impossible to know whether the H— family had been deliberately misled, or had failed to understand what they had agreed to. They seemed to have thought that colony detention would be an alternative to borstal – that of course Jessie could come home once her three years imprisonment for the crime was completed. One year after being certified, Jessie was legally entitled to a re-examination. When no news of such an event came through, Jessie's mother wrote the first of many letters to the Board of Control. Mrs H—'s agonised requests were always met with cold, curt 'no's from the Board. But she could not have been more deferential:

> I am very sorry to trouble you again but can you be so kind has to give me a little idea how long my dear daughter will be kept in Farmfield has my Husband and I was really expecting her home for Xmas and we were making a little preparation for her homecoming but if she has got to stop for another year or two we have no need to go on doing what we are. I sincerely hope I am not putting you to any trouble in answering my letter only it is quite natural I am anxious to know being her mother and I have no wish to be any trouble to you in keeping on writing hoping you will favour me with a reply it will be such a relief to me, I remain your humble servant.

She would only ever get a two-sentence letter back, the gist of which was: we can't say; probably not.

Two years into the detention, Mrs H— pointed out that what was happening in this case was indefinite detention – just for stealing items of low-value clothing. 'If you have any feelings at all you will oblige me and let me know and ease my mind how long she will be away because she does not belong to you council people. I am her mother and it is rather hard on me.'

Not until she can demonstrate that her 'mental condition' has improved, they replied.

But how could she improve in the company she was forced to keep?, her mother asked. She was not getting any stimulation when 'all the time she sees the same people over and over again . . . My daughter is amongst the Poor Girls what is worse than herself if she is kept there much longer she will be going the same give her a chance and let her come home I hope you will forgive me writing in this way.'

Darenth Colony in Kent.

For reasons that are not clear, Jessie was transferred to the Darenth Training Colony for the mentally deficient in Kent in June 1919. Here, the medical officer, who knew nothing of her case or prior diagnoses, interviewed her and informed the Board of Control: 'This patient does not present any intellectual defect as compared with other women of her age and social class... Since admission here the patient has been well-behaved and has given no trouble of any kind.'

This must have made some kind of impact as shortly afterwards the LCC did one of its snooping operations on working-class family life (always deeply resented) and noted that although he was only a casual dock labourer, Mr H—'s wages had improved and Mrs H—'s work as a cleaner meant that they had been able to move to a nicer South London street – in Kennington. They now lived in a four-roomed rented house, 'clean and comfortable', the LCC reported. Jessie's fifteen-year-old sister earned a (not good) wage at Messrs Scrubbs Ammonia Manufacturers, near the river at Vauxhall. The sister said Scrubbs were desperate for workers and she was sure she could easily secure work for Jessie at the plant. And if her daughter required a good deal of supervision, Mrs H— said she would give up her evening office-cleaning work in order to provide it.

This seemed to do the trick, and in January 1920 Jessie was discharged to the statutory supervision of her parents. Mrs H— wrote to the Board of Control: 'I shall be glad to have her I am very grateful to you for seeing into her case. I have been a bit of trouble to you in writing so much but of course it is quite natural.'

This tale doesn't have a happy ending: the lure of the ammonia factory proving weak, Jessie continued on a career of thieving. This time, though, she could not be described as a mental defective – that much had been established. And so Jessie had to do time in a proper prison. Just eleven months after getting home, she was sentenced to three months hard labour in Holloway Prison for the theft of an expensive bolt of cloth. She had given the alias 'Myrtle Hymas' upon arrest; and before long, Scotland Yard built up a file on her misdeeds, so that when later, under her new, married name, she stole two dresses from a shop, they knew it was her. The Board of Control may well have been experiencing a feeling of 'we told you so'.

11

DIRTY HOMES, STUPID KIDS

Social snobbery had an impact on who was to be labelled 'feeble-minded'. Sociologists of the 1930s and 1940s wrote books and papers on the 'social problem family' and the 'social problem group', keen to demonstrate that these were the quagmire from which the bulk of feeble-mindedness emerged. While 'idiocy' and 'imbecility' appeared scattered randomly across the social classes, with no clear hereditary cause, eugenically inclined sociologists set out to demonstrate that feeble-mindedness, by contrast, was closely linked to heredity.

The 'social problem group' was dreamt up in the mid-1920s, as the Wood Committee on Mental Deficiency investigated the likely true level of mental deficiency in Britain. The Wood Report, published in 1929, decided that ten per cent of the population were a 'social problem'; its chief investigator, Dr E.O. Lewis, an inspector at the Board of Control, examined the records supplied to him by the police, the Poor Law authorities, schools and charities because 'the names of the same families kept cropping up as presenting difficulties of various kinds. It became clear that these families constituted a problem to all the social services of the district in which they lived. The members of these families were not necessarily defective. They were, rather, retarded, low-grade normal, who were unequal to the stress of economic competition.' This was Lewis's 'social problem group', and his allegation was that they were the source of his proposed steep rise in mental deficiency – up from 4.6 per thousand in 1905 to eight per thousand in the mid-1920s. That amounted to a total of 300,000 people in England and Wales, the vast majority of them never ascertained, thus never certified – a frightening, unacknowledged human sump. Lewis claimed that twenty-five per cent of the feeble-minded lived in 'very poor' homes, and thirty-six per cent in 'poor' homes. He didn't yet know the mechanism, but he was in no doubt that his statistics proved that high-grade feeble-mindedness was transmitted biologically. To the likes of Lewis, never had the poor seemed so scary.

The social problem group was a powerful and simple concept – and the headlines wrote themselves. Technocrats and the 'social science intelligentsia' bandied the term around without ever being able confidently to define it. In his 1946 study *Families in Trouble, An Enquiry into Problem Families in Luton*, public health officer Charles Tomlinson admitted on page one, 'There is no accepted definition of a problem family . . . previous writers have emphasised the difficulty of defining the term "social problem family", and some have preferred to use other expressions, such as "derelict", "handicapped" or "unsatisfactory".' Tomlinson suggested instead the not very succinct definition: 'Those families who, for their own wellbeing or the wellbeing of others, for reasons primarily unconnected with old age, accident, misfortune, illness or pregnancy, require a substantially greater degree of supervision and help over longer periods than is usually provided by existing social services.'

Tomlinson decided that one per cent of Luton's circa 100,000 population could be placed under the 'problem' heading; as for the incidence of feeble mindedness within such families, 'the data are insufficient for definite conclusions'. He cited other recent studies, which suggested that in Rotherham there were three 'problem' families per thousand, and in Hertfordshire one per thousand, and in Liverpool 1.44 per thousand. Meanwhile in East London, E.J. Lidbetter, a prominent member of the Eugenics Society and a former Poor Law relieving officer, spent twenty-three years surveying three thousand East London recipients of welfare payments for his 1933 work *Heredity and the Social Problem Group* and satisfied himself that 'there is some evidence that the persons included in the pedigrees have a sufficiency of common characteristics such as to constitute a class by themselves'. Lidbetter believed there were two subgroups within the welfare recipients who were suffering conditions that were passed on from their forebears: those with obvious 'defects', including insanity and epilepsy; and a second group of 'the merely low-grade type'. He drew elaborate diagrams to show they often intermingled 'and it is probable that the latter group is the recruiting ground for the former'. Without deigning to supply any evidence, he stated 'the transmission of defectiveness ... occurs with remarkable frequency.' *Heredity and the Social Problem Group* was to have been the first in a multi-volume survey to prove the inheritance of a number of mental and physical handicaps among recipients of welfare relief; so poor was its argument, it was shelved after this first volume.

In all of these interwar surveys, inspired by the Wood Report, hard data on causation were wholly lacking. One of the ways to mitigate the glaring lack of conclusive evidence (a trick, if you like) was to remind the reader that mental defectiveness could be hidden – that parents or relatives who couldn't obviously be labelled as defective were nevertheless hiding their abnormality behind a façade; inside, their genetic material was awash with undesirable characteristics. This is an echo of Mary Dendy's 1910 worry (mentioned in chapter 5) concerning the 'well veneered' who could pass for 'almost normal' – the people 'against which there is most need to protect society'.

In Luton, Charles Tomlinson discovered lives sent spinning off track by domestic violence, divorce or separation, frequent pregnancies, heavy drinking, and the absence of a male breadwinner plunging the family into chronic poverty. Mrs D— had had to flee her violent husband with her four children; Mr and Mrs E— had come down from Clydeside and separated; Mrs G— had had a brutal stepfather who threw her out and while still very young she married a soldier who was 'slightly backward' and proved unable to provide for her and their child; Mr J— was in and out of prison and could not provide for his wife and son.

Tomlinson was much more fascinated and perturbed by their dirtiness and untidiness – both in their person and in their homes – than the interaction of family breakdown and domestic violence with chronic poverty and lack of agency. His survey catalogued the phantasmagoric squalor of the Luton families he had labelled 'problem', and he quoted at length a notorious passage from the article 'Problem Families' in *Public Health* magazine, September 1944:

> From their appearance they are strangers to soap and water, to toothbrush and comb; the clothing is dirty and torn, and the footwear absent or totally

Charles Tomlinson included these pictures of conditions in parts of Luton in his 1946 book *Families in Trouble*. He wanted to explore the extent to which heredity and environmental conditions impacted on IQ and aspiration.

> inadequate. Often they are verminous and have scabies and impetigo . . . The mother is frequently substandard mentally. The home, if indeed it can be termed as such, has usually the most striking characteristics. Nauseating odours assail one's nostrils on entry, and the source is usually located in some urine-sodden, faecal-stained mattress in an upstairs room. There are no floor coverings, no decorations on the walls, except perhaps the scribblings of the children and bizarre patterns formed by absent plaster.

Mothers were blamed far more than fathers for the creation of a problem family in the various interwar and postwar surveys. Tomlinson wrote that 'as might be expected . . . a poor-type mother is usually a more serious matter for the family than an indifferent or subnormal father.' An incompetent, stupid or lazy wife and mother caused her husband to go on long absences, to turn to drink, womanising, gambling, and yes, even to resort to violence against her – this view crops up repeatedly in the surveys. But if a mother decided to go out to work, in order to supplement, or even to fully replace, the father's wage, that too could be laid at her door for causing neglect to the children and to the running of the home.

Apparently without realising it, Tomlinson had provided an environmentalist, not eugenic, explanation for the 'problem'. There may have seemed some correlation between squalor and mental deficiency, but he failed utterly to show causation. This was unfortunate, as *Families in Trouble* had been part-funded by the Eugenics Society. This was 1946, and eugenics was struggling to regain its own respectability following the revelations of the Nazi atrocities; but here was the Eugenics Society attempting to get up a case against the chronically impoverished. It was as though the 'discipline' hadn't moved on for fifty years – still no hard evidence on the fertility of the 'problem' population; still no reliable family histories, or 'pedigrees', as they queasily called them.

Their new(ish) line of attack was to attempt to prove that the measures put in place to improve the lives of the chronically poor had failed to 'solve' the problem of the alleged ten per cent at the bottom of the social scale. The vast rise in new-build decent affordable homes by local authorities; the army of health visitors and social workers intervening to offer targeted assistance; the introduction of family allowances (after 1945) – it was gleefully pointed out by social conservatives (opposed to public spending on combating the effects of poverty) that these had barely touched the ten percenters. *They* continued to wallow in conditions that were as bad as anything in the nineteenth-century slum: 'A slum area has been cleared and the population moved to well-designed houses and to generally hygienic surroundings, [but] the very conditions which it was hoped to remedy soon recur and oblige the local authority to intervene to prevent the emergence of a new slum area', wrote E.O. Lewis. 'It was in such a district as this that our Investigator found one of the highest incidences of mental defect found in any area of comparable size.'

In 1943 the study *Our Towns: A Close-Up* was published by the Hygiene Committee of the Women's Group on Public Welfare, a group that had Labour MP Margaret Bondfield as its chair. As evacuated children and (frequently, in the first wave) their mothers came *en masse* into rural areas, many villagers expressed shock and disgust at their physical condition and behaviour. The urban slum and its subcultures became the focus of national alarm. Chapter 2 of *Our Towns* is entitled 'Living Below Standard' and its sub-headings are: Wrong Spending; Bad Sleeping Habits; Bad Feeding Habits; Juvenile Delinquency and Want of Discipline; Dirty and Inadequate Clothing; Skin Diseases; Insanitary Habits; Bodily Dirtiness. Bondfield wrote that the evacuation phenomenon offered a rare opportunity to undertake a truly national survey of urban life, since all major conurbations were represented in the data. Bondfield (firmly of the environmentalist, and not eugenic, view) hoped that the findings would stimulate ideas for the best way to reconstruct Britain after the war – how most efficiently to eradicate the slum conditions that were at that point causing so much shock to countryfolk. 'The effect of the evacuation was to flood the dark places with light and bring home to the national consciousness that the "submerged tenth" . . . still exists in our towns, like a hidden sore, poor, dirty, and crude in its habits, an intolerable and degrading burden to decent people forced by poverty to neighbour with it.' In the event, the Attlee government of 1945–51 did push through the reforms Bondfield wanted to see, including a national health service, family allowances and nursery education.

Our Towns presents a horrifying picture of lack of national cohesion – a smack in the face of the cosy Home Front propaganda pumped out to keep morale high; its contents very much go against the grain of plucky

Britain united against the common enemy of Nazi Germany. Bondfield, making an effort to be fair, also included evacuees' complaints of the brutality shown by some of the rural hosts towards the youngsters billeted with them, as well as more general unkindness, plus the exploitation of their labour. Perhaps what is most interesting is that nowhere in *Our Towns* is any space given to the idea that the alleged moral, or behavioural, failings of urban children and their mothers were inherited: terrible conditions and the formation of deeply ingrained habits and attitudes were the culprit.

Charles Tomlinson, on the Luton Home Front, was of the view that the type of state interventions championed by Bondfield and those who thought like her into the lives of the working classes had in fact acted to sharpen and separate off the ne'er-do-wells from the 'respectable' majority: 'Public opinion is less tolerant of persistent lousiness, localised squalor, and ill-clad, neglected children', he wrote. 'There is a growing demand, based on sheer self-defence, for more extensive powers to aid and control the few who cannot or will not help and discipline themselves.'

If Tomlinson was correct, it suggests that the working class itself would increasingly welcome harder-line policies against the outcasts of society – that they were keen to avoid the taint of being associated with scum. It's an interesting point, but not original to the mid-1940s. The investigations of London working-class life undertaken by Charles Booth between 1886 and 1903 posited a definite divide between the self-identified 'respectable' working class and the 'loafer' class. Indeed, during the 'New Unionism' of 1889–95 certain trades union leaders similarly sectioned off the unproductive and seemingly parasitical poor from the sturdy, hardworking majority seeking better pay and conditions. And wasn't this Will Crooks's very point in his 'vermin' speech to parliament in 1912 (quoted in chapter 7)? The great social investigator Seebohm Rowntree (who devised his famous sociological survey of York in 1899) had himself become a pessimist by 1946. He wrote that

> if the nation is seriously concerned to abolish poverty and the slums, it will no longer be able to avoid facing the problem of the subnormal family. As the economic level of the poorest classes is raised and their standards of welfare are improved, the problem families stand out more clearly as a

minority who do not benefit from the improved conditions, but remain a menace and a disgrace to the community.

People who thought this way could even call on Karl Marx's and Friedrich Engels's term coined in the 1840s, 'the lumpenproletariat' – consisting of tramps, beggars, prostitutes, thieves, incapable of class consciousness and class solidarity, and therefore counter-revolutionaries.

To what extent was 'pride' in being a member of the 'respectable working class' an expression of internalised self-hatred? Could it be that the working class, in the face of universal contempt and shaming, were understandably giving in to what historian John Macnicol has described as 'the fetishisation of middle-class mores'? Had decades of church, school and philanthropic intervention into their lives eroded a sense of common grievance regarding social inequality, offering in its place the urge to get up and out of a class, leaving the rest to rot? Engels again – he popularised the concept of *embourgeoisement*. The factors that could cause fluidity between what was deemed 'respectable' and 'unrespectable' would rarely be analysed until the 1950s, with few social investigators mentioning such factors as childhood trauma, brain injury, febrile illness, being the victim of violence or of sustained psychological bullying, or of long-enduring chronic poverty (with deprivation stretching back through the generations) in creating an anti-social personality and 'social inefficiency'.

One of the exceptions was geneticist and psychiatrist Lionel Penrose, who undertook the last great study of mental deficiency and heredity before the Second World War. His 1938 'Colchester Survey' of 1,280 patients at the Royal Eastern Counties Institution who had intellectual disability challenged the views of many of his fellow geneticists. He discovered far greater complexity in the likely causes of a wide variety of mental conditions. Penrose didn't even agree that a 'social problem group' existed.

He had studied psychoanalysis in his early career, and brought an understanding of child psychology to his work. In 1945 he would become Galton Professor of Eugenics at University College London. Nevertheless, his opinions were most unGaltonian:

> It is not usually believed by those who study mental deficiency that the retardation observed can be due to psychological causes ... In view of the

recent work of psychologists who have stressed the exceptional importance of the first few years of life in the formation of character, it is well to reconsider this belief... It is hardly necessary to point out that the educational opportunities of different children vary with social class. Lack of such opportunity may, later on, cause failure in mental tests of a scholastic nature.

Penrose had no doubt that many mental deficiency certifications were less to do with the genuine ineducability of a youngster than with 'extraneous circumstances', including family background. In addition, and despite the Act of 1913 expressly differentiating between 'the merely dull and backward' (who could be improved by attending a special school) and those with 'permanent mental defect', all too often that crucial distinction was blurred. Penrose found that at age twelve, thirty people per thousand fell within the range of mentally defective; but this fell to 8.4 per thousand in the age-range twenty to twenty-nine. This meant that many 'abnormal environmental conditions', which were reducing educable capacity during adolescence, had disappeared by the time a person reached young adulthood. It was therefore dangerous to impose lifelong detention orders on people between the ages of twelve and eighteen – when their educability and 'social efficiency' were so uncertain, so in flux. It also torpedoed the claims that IQ was overwhelmingly inherited and fixed, for life, within a limited range.

Scepticism regarding mental testing was as old as the tests themselves. In 1929 G.K. Chesterton had cited cultural differences as having a huge impact on scores. Writing in response to the Wood Report finding that rural areas showed vastly higher levels of mental deficiency than urban, Chesterton wrote to the Minister of Health:

The rural mind has been taught caution in a hard school. No countryman gives his confidence to a stranger, and his children share this distrust. Much more than urban children, rural children would be shy, distrustful, inarticulate or sullen before a stranger who demanded purposeless tasks alien to the realist, rude mind. Rural speech, which is considered by townsmen vague, is not unintelligent, is really highly allusive. To his fellows the rustic's speech has a precision which escapes a townsman. The investigator might, perhaps, have remembered the familiar saying, 'If you want to find a fool in the country, take him with you.'

Chesterton believed that the Wood Committee had overlooked such environmental factors as malnutrition, bad housing and insecure work prospects: 'This applies in a measure to farm labourers as well as to slum dwellers.' Lewis put the mental deficiency rate in urban slum areas at 6.71 per thousand but a staggering 10.49 in rural areas. (This, of course, contrasts with the later *Our Towns* findings, which told a story of a socially inefficient and filthy urban underclass contaminating a sturdy, civilised rural population.) It must be pointed out that no other survey came close to Lewis's statistics, and in later decades, his research would be seen as an overstatement of his claim that mental deficiency levels had soared in the first thirty years of the century.

The Report of the Wood Committee attempted to explain the rural–urban difference as the result of the younger, more energetic and ambitious (the 'socially efficient') countryfolk migrating to the towns and cities – leaving their less sharp, dozier neighbours to swell the ranks of Dr Lewis's tables. The Report didn't use the word 'inbred', but suggested that this internal migration 'has left behind a population inferior in mental quality . . . and that the inter-marriage of this inferior stock has produced a larger number of mental defectives during the last fifty years than in previous centuries.'

For his Wood Committee work, E.O. Lewis had used the Binet–Simon intelligence tests (which G.K. Chesterton described as 'unsound and mischievous', suiting 'some types of mind better than others, which are not therefore inferior'). Early mental testing, such as that devised by Francis Galton, had measured basic skills and sensory reactions. Between 1905 and 1911, Alfred Binet and Théodore Simon tested hundreds of Parisian schoolchildren, suspected of being mentally defective, on their spatial perception, deductive skills, range of vocabulary, memory, judgment and 'moral sense'. They devised their hugely influential concept of 'mental age', and the 'Binet–Simon method', along with variations upon it, became crucial in twentieth-century decisions about whether a child or adult was 'merely dull and backward' or 'mentally defective'.

Chesterton said that forty-seven of the 112 questions in the IQ test used by Lewis were arithmetical, and that 'arithmetic is not a test of IQ'. He added that some of the rubric on the picture tests was incorrect; and one of the pictures was truly meaningful only to someone who understood American culture. Moreover, the tests concerning pounds, shillings and pence were unfair to rural children, who would be far less likely to even see a shop regularly, or to handle cash in doing errands for their parents.

As historian Mark Jackson has pointed out, these early IQ tests confirmed a spectrum, rather than sharp divisions in intellectual capacity, 'stressing the continuous distribution' of intelligence within a population. Jackson quotes Karl Pearson as stating, 'In truth, there is no such intellectual boundary between the normal and the mentally defective. The distribution of intelligence in both normal and defective is absolutely continuous and in the purely psychological tests of intelligence there are no rigid and limited categories of normal intelligence and feeble mindedness.'

More mundanely, regardless of the outcome of an IQ test, decisions about which children to certify as mentally defective and send to a colony, and which to educate in a special school, were often made simply according to the availability of spaces at a local special school. If there were none, the mental deficiency colony beckoned. A child's fate really could – and too often did – pivot on where in the system any slack was to be found.

Alfred Binet had wished to make clear that the Binet–Simon tests should be just one tool used in the analysis of a child: 'Our examination of intelligence cannot take account of all those qualities, attention, will, regularity, perseverance, teachableness and courage, which play so important a part in school work and also in after life.' He also introduced word-association tests, asking a child to say as many disconnected words as s/he could in three minutes. Separately, the Porteus Maze was devised in 1914 – this comprised a labyrinth on paper through which the path in and out, or in and to the centre, must be traced in pencil; it was believed to uncover the tested person's level of recklessness and impulsiveness.

From 1918 the Jungian psychoanalytic school sought out repressed emotional complexes by compiling a list of words likely to provoke an emotional reaction, or to trigger memory; the tested person was to call out the first word that came into their head as each was read out. Next, the 'psycho-galvanic' method was discovered – with electrodes reacting to changes in the skin caused by an emotional reaction (or no reaction, in the case of psychopaths). Further attempts to gauge a child's moral sense included sets of pictures that the subject had to rank in terms of personal preference. In the United States, a child was asked to arrange a list of criminal offences in order of wickedness.

Half the British Boards of Education resisted the introduction of IQ testing, with many teachers disliking and distrusting the phenomenon. The Assistant Secondary Schoolmasters conference in January 1950 called for the abolition of IQ tests. One master from a Dorset grammar school said, 'They have a pseudo-scientific appearance of infallibility and impartiality but are comparatively useless. They put a premium on slickness to the disadvantage of the painstaking, if slow, worker.' The very conditions under which IQ was tested could be distressing for a sensitive child, and one teacher recalled, in 1949: 'I myself have been present when tests have been given without the slightest spark of any emotion at all, when the tester has sat gazing at a stop-watch, only looking up to pull apart the jigsaw . . . which the child has so laboriously constructed, and in the coldest of voices to ask for it to be done over again.' This teacher also spotted the phenomenon of 'the neurotic child, whose mental energy is so taken up with internal conflict that he has little to spare for the retention of material which requires a concentrated span of attention'.

Meanwhile, one Mrs C— revealed in 1949 that during testing, her son had failed to complete the test on time (though all his responses were correct) and was therefore diagnosed as mentally defective. But after seven weeks of intensive residential tuition with a retired teacher, the boy was able to read and speak more fluently, and upon re-testing, dodged the fate that the local authority had lined up for him – certification into a mental deficiency colony.

Many mothers expressed similar concerns about IQ testing to Judy Fryd, the founder of the National Association of Parents of Backward Children (today's Mencap). The contempt shown to such women was clear, as Fryd wrote: 'Our fears [about IQ testing] were dismissed as the unreasoning resistance of Doting Mammas.' She continued: 'I believe there are many clever and competent children diagnosed as low-grade mental defectives who may one day be enabled to use their powers in a socially acceptable way; this can only come through special educational facilities being made available and through further researches into brain surgery and neuro-therapy."*

Back-chat while being tested could be fatal. Sixteen-year-old Noele Arden came before a mental deficiency panel in the late 1940s. To test her intelligence, they asked her to explain the difference between an orange and a lemon. 'Suck it and see', she replied (correctly). She was sent to Rampton, and stayed nine years.

The British public appear to have been split in their opinion of the validity of IQ testing. On the one hand, after the Second World War, IQ tests published in cheap paperback book form sold very well, and there was a vague general belief that IQ mattered – that in some way it was linked to respectability and social usefulness. But on the other, popular antagonism to the nature of the tests (and to the 'experts' who set them) was easily aroused. The *Nottingham Evening Post* received a huge cache of readers' letters when it published a brief court report in its news pages, on 8 January 1949. 'I am not a mental defective', thirty-seven-year-old James Smith told the magistrate when he came before the court for breaching his probation.

> I have never known true happiness and a home. I was born in the slums of Stepney. My father was a drunkard and used to ill treat us. My mother protected us as much as she could. I lost her in a blitz on the East End of London . . . I work [as a painter] from 7.30 a.m. to 9.30 p.m. and have been complimented on my work. Please give me a chance to prove my worth and carry on my occupation.

Part of Smith's probation terms had included treatment at Saxondale Psychiatric Hospital; here, the chief psychiatrist administered an IQ test, in which Smith performed badly. The authorities now wanted Smith certified as a mental defective and confined permanently. The Saxondale

---

* When Judy Fryd (1909–2000) found little official support for her learning-disabled daughter, she wrote, under the pseudonym 'Cinderella', to *Nursery World Magazine* in 1946 asking other parents if they were having the same experience. *Nursery World* received over one thousand responses, and so Fryd set up her organisation.

psychiatrist told the magistrate's hearing that Smith could not tell him how many three halfpennies there were in a pound; how many pounds in a ton; or how to spell the word 'religion'. 'He has a superficial glibness and no sense of social responsibility', the doctor said. 'I am fighting for my liberty and freedom', replied Smith. Three days later, the *Post*'s 'Editor's Letter Bag' column was filled with sympathetic correspondence.

> 'I read the sad story of James Smith with a very troubled mind indeed. I venture to say that thousands of people would have failed the doctor's tests.'

> 'Many people who are competent when facing an examiner, through sheer nervous reaction, blunder hopelessly.'

> 'I know of several sensible people who could not answer that pounds, shillings and pence question without doing a mental sum. What is the "superficial glibness" as applied to Smith? Is it an ability to argue a case with coherence and some degree of skill? The whole affair appears to have been most unsatisfactory. It is essential that the liberties of all, whether of high intellect or limited mental faculties, should be jealously protected.'

The editor, however, was not convinced. In his column alongside the letters, he expressed approval that the ordinary British citizen still rushed to the defence of individual liberty when it appeared to be under threat. However, he continued, he had sought out the opinion of a psychologist regarding Smith's failure in the test; this psychologist had pointed out that it wasn't so much the quality of the answers as the attitude to being tested that had marked Smith out as feeble-minded. That was the true secret of IQ testing, he revealed. The editor paraphrased the psychologist as stating that

> the ordinary person will either attempt to work out the problem mentally, or will say something like, 'Well, I'm not very good at arithmetic.' If a snap answer is given to such questions, it is held to denote lack of self-control that in some circumstances would mean danger to others. Many other questions may be asked to ascertain the true state of the subject's

mind, and it may be found that while the body has matured, the mind has remained at childhood stage.

There was something in this. When Binet and Simon had devised their testing, at the start of the century, *how* an individual attempted a question formed part of the assessment. The morally-mentally defective were those who lacked honesty, maturity and staying power. The emotionally unstable, those with an immature temperament and over-developed ego, would tend to argue during the test with the person who was administering it, or would give up the test after making very little effort. This was the 'morality' or 'social efficiency' aspect of the Binet–Simon tests. Binet–Simon sought to discover the ability to maintain thought in one definite direction and the capacity to adapt mentally in order to achieve a desired end. Furthermore, the power of self-criticism (admitting, for example, 'I've never been very good at sums') also indicated maturity and stability – even though the correct answer was not in fact achieved. Self-control and honesty also could be tested by asking, for example, for a pencil tracing to be completed while a blindfold was worn; peeping indicated a tendency towards criminality.

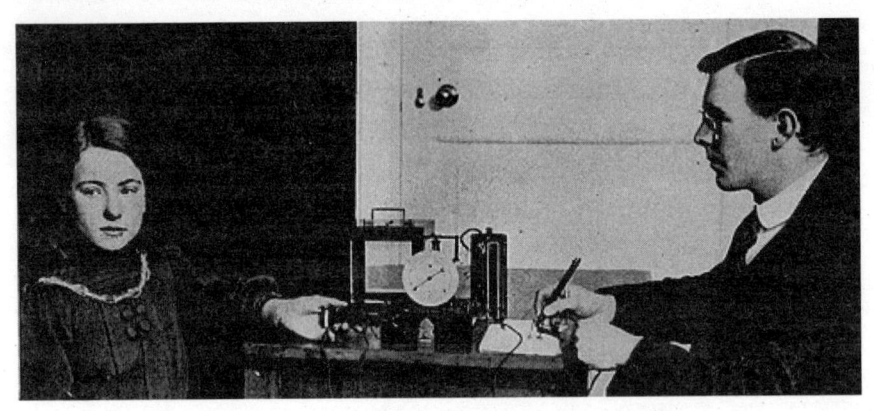

Cyril Burt attempting to measure thought at the
London County Council in the 1920s.

Cyril Burt, the pioneering psychologist introduced in chapter 9, believed that IQ was overwhelmingly inherited and that this inheritance decided the socio-economic course of an individual's life. In his 1943 article 'Ability and Income', he tabulated his findings in this way.

		Child IQ	Adult IQ
Class I	Higher professional: administrative	120.3	153.2
Class II	Lower professional: technical/executive	114.6	132.3
Class III	Highly skilled: clerical	109.7	117.1
Class IV	Skilled	104.5	108.6
Class V	Semi-skilled	98.2	97.5
Class VI	Unskilled	92	86.8
Class VII	Casual	89.1	81.6
Class VIII	Institutionalised	67.2	57.3

Shortly after his death in 1971 the first of several claims were made that Burt had falsified his data in order to waterproof his case that heredity trumped any environmental factors regarding IQ and social class. The present-day consensus is that his post-Second World War work was fraudulent in this regard. Burt had found his beliefs under fire from certain 'reform eugenicists': as historian Daniel Kevles has revealed, biologist Lancelot Hogben and his Social Biology Group laboratory at the London School of Economics set out in the 1930s to debunk the IQ test as proving anything other than being good at taking tests. Hogben et al. built up a large dataset and found that a significant proportion of children between the ages of nine and twelve, with IQ results of around the 130 mark placing them in the 'gifted' or 'talented' range, came from low-income families; the Social Biology Group thought it was iniquitous that such children would have little chance of being able to stay on at school beyond age fourteen, let alone progress to tertiary education. (This was the fate of my own working-class mother – tested in the early 1940s and scoring highly, but with no chance to progress beyond the local board school.)

There was one other group that tended to score quite highly in IQ tests – the person who was very good indeed at being a criminal. As Daniel Kevles revealed, in the United States, IQ tests showed criminals did at least as well as members of the US Army. American psychologist Carl Murchison stated that the characteristics that 'make for worldly success in business or professional life also make for success in crime'.

Cunning, craftiness – on the one hand these could be said to indicate moral squalor; on the other, superior intelligence that could raise one out of the slum, by either fair means or foul.

12

LOVESICK IN ST JAMES'S

As far away from a slum as could be, John S— perturbed his wealthy widowed mother to such an extent that she called in the Board of Control. She and John divided their time between a mansion in Essex and a pair of smart metropolitan flats in St James's Court Mansions, Buckingham Gate. She was at flat number 183; John a couple of floors below at number 73.

Twenty-year-old John was proudly, unashamedly gay – a wonderful, courageous stance, yes; but unwise in 1915. The Board of Control and the doctors consulted by his mother were horrified, and morbidly curious, about his passion for men. They were also acutely aware that, here, they were dealing with a young *gentleman* – this highly unusual intervention by the Board into the lives of the super-wealthy would mean they must take extra special care.

For the most part, upper- and upper-middle-class families were able to deal discreetly with their severely learning-disabled family members. Famously, Queen Elizabeth II's first cousins, Nerissa and Katherine Bowes-Lyon, were quietly placed in Earlswood Hospital for the mentally disabled, in Surrey, in 1941, when each was in her early twenties. They were described in the horrible terminology of the day as 'imbeciles'; and their existence was erased by the printing of false death dates in *Burke's Peerage*. For those wealthy people deemed only to be 'feeble-minded', they could have a close eye kept on them without being placed in an institution – out of sight, at home with salaried tutors, governesses, companions, valets and other senior domestic staff. All of which, as suggested in earlier chapters, helped to promote the (erroneous) idea that mental deficiency was a problem disproportionately affecting the less-well-off.

John S—'s case is, therefore, unusual, and it made the Board of Control nervous. Here were a bunch of middle-class bureaucrats being insufferably nosy about the affairs of an heir to a fortune; yet they had been requested to be so by his concerned mother, who had sought, and was granted, statutory guardianship. The tension is apparent in the letter sent by Mrs S—'s lawyer, Sir Charles Russell (baronet son of an even more prestigious father), who wrote the following to the Board, in January 1917:

> From time to time you send a Visitor to call to see the young man, and of course, Mrs S— is only too glad to do all she can to assist him. She has, however, been rendered very unhappy by the gentleman's mode of announcing himself as 'A Visitor from the Board of Control, come to examine Mr S—'. This to the servants at St James's Court Mansions. You will readily understand that such a thoughtless procedure causes considerable humiliation and annoyance to the parties concerned. Could you be good enough tactfully to suggest to the gentleman in question that it would perhaps be well for him simply to send in his card and ask for Mrs S—, Mr S—, or Mr Costigan, his tutor?

Costigan was about to leave his post, in order to get married. He had been John's tutor for five years, and it was this change in John's circumstances that had made up his mother's mind, in 1915, to apply for John's certification as a mental defective, of the moral imbecile variety, under the recent Mental Deficiency Act. There was never any question of John being institutionalised; she simply wanted the legal power, under guardianship, to compel him to stay by her side so that she could prevent him approaching men for sex. When, in January 1917, John started agitating to be allowed to go and stay for three or four months at the Piccadilly Hotel 'for a change, because he likes the food there', his mother was able to use his certification to make it clear that he was not a free agent.

His mother also knew that when John came into his fortune at the age of twenty-one, he would be vulnerable to being fleeced of his money. She had noted that his crushes on various men led him to spend extravagantly on them. Worse still, his homosexuality made him vulnerable to blackmailers. Under certification, his income and his property were fully protected, and all adventurers thwarted.

John's effeminacy enraged the family doctor, Cecil Christopherson, who was, in 1915, a captain in the Army Medical Corps, a family man, a golf fanatic. The nation's young men were sacrificing themselves at the Front, but here was John being

> pronouncedly feminine in his manner and in his talk. He is very fond of needlework. He powders himself, uses a great deal of scent and spends hours dressing himself with colour schemes etc ... I believe he is a congenital homosexual pervert and I do not believe there is any hope of his recovery. In my opinion, he is a feeble-minded person in whose case from birth or from early age has existed mental defectiveness.

Equally horrified was the Board of Control 'Visitor' (the one whose insensitivity had annoyed Mrs S—). He reported that

> it was about noon when I called to see Mr S— today at his flat in St James's Court Mansions. I found him cleaning his bedroom, 'doing housemaid's work', as he expressed it, attired in a sleeping suit and a very ragged dressing gown. He had the appearance of not having shaved, and his hair was untidy. There is no doubt as to the luxury of his surroundings, or generally as to the satisfactory nature of his material welfare. But it seems to me, taking into consideration his known proclivities ... that those who have care of him should at least make some effort to induce him to adopt regular habits and observe the ordinary decencies of life.

John spoke openly to Dr Christopherson about the sexual abuse he had experienced from a maternal uncle, and the doctor put it in these words: 'In February 1911 I was told that he had misconducted himself with his uncle, and I had a conversation with him on the matter. He did not express any horror whatever about his conduct or that of his uncle, and I could not get him to express any regret or sorrow ... for he was not in any way frightened, surprised or disgusted at what had taken place, nor did he in any way resent what had occurred.'

No action was taken against the uncle, whose own homosexuality was cited in the long 'pedigree' that Christopherson, plus a second doctor and the Board of Control, produced to show that John's mental condition was inherited. ('His family history is on both sides distinctly bad.') One of the problems eugenicists had found in trying to prove the inherited nature of mental deficiency was that the bulk of the population had no documented lineage – 'case histories' foundered all too often, with even parentage being uncertain, let alone grand-parentage and beyond. However, the further up the social scale, the better known the fate of ancestors, and so John S— was partially certified on the grounds that his maternal uncle was a homosexual; his aunt and three first cousins of his father were epileptic; his maternal great uncle 'was a drug maniac'; his first cousin was a dipsomaniac, while another first cousin was 'mentally defective with vicious tendencies'. John had had an epileptic fit at age thirteen (in a train carriage, London to St Leonard's, right in front of Dr Christopherson) and occasional bouts of petit mal (absence seizure) subsequently. The second certifying doctor, C.G. Drummond Morier, additionally pointed out that 'he has a peculiarly shaped head, which is larger on one side than on the other'.

In seeking any other reason that John had turned out as he had, the doctors wanted to part-blame a boy at his boarding school. 'It has been proved beyond all question that he [John] is the subject of masturbation. This habit may have been commenced in him through the evil influence of another boy at a school where he was.' Dr Christopherson circumcised him in the hope this would stop the self-abuse, 'but as soon as he was soundly healed he told me that he had reverted to the practice'.

Next Dr Christopherson placed him, aged sixteen, under the residential care of neurologist, psychotherapist and twice struck-off pioneer of experimental medicine Dr Hadyn Brown. He stayed at Dr Brown's home in Caterham, Surrey, for several months, and 'this benefited him for several weeks, but he told me that it had in no way lessened his desire for men. He told me that his desire was to act the passive part rather than the active in his connection with men. He told me that whenever he met a well-formed man in the street it immediately aroused a sexual desire in him.'

A series of infatuations laid John low, and Dr Christopherson was alarmed when John stopped eating and sleeping during one particular incident of unrequited love. He 'in every way behaved like a boy who is lovesick ... He was in the habit of surrounding this man's photograph with flowers, in his room.'

In the early summer of 1910, when he was sixteen, Dr Christopherson noted that 'he had contracted a violent attachment for a skating rink instructor, and I spoke to him on the matter. He could apparently think and talk of nothing else. He told me he was attracted by the man's figure.' Why didn't John feel shame?! Dr Christopherson was amazed: 'He discusses the matter in a most open manner, and appears to think nothing of it. Indeed, he is boastful on the subject, and expresses himself as quite regardless of what the world in general thinks on such matters ... No arguments appear to be capable of inducing him to recognise that his homosexual tendencies are wrong; he treats all such arguments with contempt.'

All homosexual acts – even between consenting adult men, even in private – were illegal, under the infamous 'Labouchère Amendment' of the 1885 Criminal Law Amendment Act; this clause was lobbed in late at night from the back benches, in a sparsely populated House of Commons, by journalist and rabidly anti-homosexual Liberal MP Henry Labouchère. He had tacked it on without notice to an Act that offered greater protection to women and girls from sexual assault. After 1885, and quite separately to legislation against sodomy/ buggery, any behaviour between males that came under the vague umbrella term 'gross indecency' was punishable by up to two years in prison, with or without hard labour. If the acts took place in open spaces, such as parks and public gardens, they could be punishable by a financial penalty under the 1872 Parks Regulation Act. In addition to the judicial punishment, a man's reputation was wrecked for ever. This is why blackmailers targeted any male against whom a story could be got up that he indulged in sex acts with men, or importuned others to commit such acts. The Labouchère Amendment's nickname was the Blackmailer's Charter.

Legal historian George Ives described the phenomenon in these words, in 1914: 'A number of most villainous gangs are always wanted by the state ... Certain blackmailers are almost as "known" as politicians or actors, only the witnesses and victims will not come forward and the police cannot get legal evidence enough to put before a jury.'

Gay men in London needed to be wary not just of blackmailers and the Metropolitan Police but of volunteers of such organisations as the National Vigilance Association, the Public Morality Council and the Central South London Free Church Council – all intent on 'purging sexual dissidence from public space', as historian Matt Houlbrook has put it.

John's homosexuality was not enough per se to see him officially branded feeble-minded: it was his unshakeable insouciance – his refusal to accept that any consequences were likely to follow on from his behaviour – that made the doctors' job so much easier. His *attitude* towards his homosexuality was unacceptable. As with the false allegations of 'lunacy' of the nineteenth century, any person who either did not comprehend, or refused to acknowledge publicly, the problematic nature of their sexuality in an aggressively heterosexual century was in danger of having this worked up into an allegation that they were 'of unsound mind'. Under the 1913 mental deficiency legislation, John's repeated refusal to compromise on his views and his intentions made him a 'moral imbecile'.

(It was in the psychiatric, rather than mental deficiency, system that atrocities against gay men would get under way, in the early 1950s. Historian Tommy Dickinson has revealed the 'aversion therapy' torture regime that was trialled at certain British mental hospitals. 'Treatments' included electric shocks, emetics to induce vomiting, and sleep deprivation. 'I remember feeling sickened by what we did to him, and it still haunts me to this day. I was a coward and selfish', one of the male nurses who administered aversion therapy told Dickinson.)

The Board of Control knew that John did not lack intelligence; in fact, they noted his knowledge of fine art and the decorative arts and his excellence as a pianist. But the doctors and the Board honed in on what they considered his shallowness of mind – his refusal to set himself to the serious study of any subject. He would not 'apply his mind to real study. All the literature that he is interested in is on sexual matters of various kinds. He becomes quite bored by any conversation on serious subjects.'

The small S— household was full of intrigue, with the tutor reporting secretly to the Board of Control that John was plotting with the valet to obtain greater freedom, but that the chauffeur had learned of this from the valet and told the housekeeper who had told the tutor. Dr Morier attributed John's scheming and plotting to craftiness and dishonesty. Morier reported that 'he rarely looks one straight in the face and there is a look about his eyes which suggests that he is cunning. His whole manner is peculiar, and his way of speaking suggests great conceit.'

The Board even put a stop to his volunteering work near the Aldwych, among the Belgian refugees who had fled the fighting, suspecting he was using this as cover for seeking out men.

The Board was astonished to learn that a recruiting officer had turned up at St James's Court Mansions in November 1915, insisting that mental deficiency was not a sufficient excuse for not enlisting to fight. In fact, the 1915 Registration Act explicitly exempted mental defectives from signing up to the Army. The Board of Control instructed the officer not to attempt this again.

When the War ended, Mrs S— appears to have had a change of heart. Although anxious, she did not try to hamper John when he began flitting between the Piccadilly Hotel, the Hyde Park Hotel, Claridge's, an apartment in Rutland Court, Knightsbridge, and another in Basil Street, Knightsbridge. He was in the company of Costigan's replacement, a Mr Hargreaves, described more frequently in the paperwork as 'companion', rather than 'tutor'. The Board of Control was also uneasy about this, but made no effort to censure John or his mother. In early 1919 she wrote and told them (she didn't *ask* them) that she, John and Mr Hargreaves were about to travel abroad together – to the Riviera ('as he does hate the cold of winter'), then on to Capri. In 1921 they headed to Paris, Egypt and Palestine.

The Board decided that over the preceding six years, John had become 'more manly ... He seems to be somewhat more composed in manner and though still physically full of restless movements, he did not jump about from one subject of conversation to another as he did before ... He told me that he has quite overcome his homosexual tendencies.' Mrs S— informed this inspector that his 'tendencies' had waned but that she had 'reason to believe that he is occasionally given to self-abuse'. She continued, 'There has been no recurrence of these [homosexual] tendencies, or even a hankering after them, for some considerable time, and I think that now the bare idea of such practices is repellent to him ... He has promised to put his capital in trust when released.'

In the winter of 1921 John and Hargreaves travelled together across America. Mrs S— asked the Board of Control if John could be released from his certification. Without a murmur, they consented. In March 1922 he was a free man, and in August of that year he set sail for New York.

It is difficult to see what had really changed. John will still have been attracted to men. He was still vulnerable to blackmailers, and to prosecution if he were to be caught in an act of 'gross indecency'. Was it simply the case that Mrs S— had proved such a diligent custodian of her son that the Board felt able to relinquish oversight? They had clearly never felt comfortable policing someone from that echelon of society. Or was it that Mrs S— had become afraid that if the guardianship continued, her relationship with John would be ruined for ever? This statement of hers suggests that may have been the case: 'He has never borne me any ill will or shown any vindictive spirit at any time since he was placed under control. It is because I am anxious to continue this affection and unbroken trust and to retain my influence over him that I am applying for his earlier release, because I feel that by making this concession and giving him an instance of my good will, he would be most likely to reciprocate it.'

John died at the age of forty, cause unknown. He had been dividing his time between a small castle in Cap d'Ail, on the Riviera, and a mansion flat in Queen's Club Gardens, West London. He left his fortune of £55,000 to his mother (equivalent to £4.7 million today).

***

Most gay men and boys who found themselves labelled as moral imbeciles after 1913 entered the mental deficiency system in quite a different way to John. They were certified under the 1913 Act via its Section 8 – that is, during prosecution for, or after being convicted of, an offence. Often this was because they had committed, or attempted, non-consensual homosexual activity – a 'sex crime', as we today would call it. However, because of the Labouchère Amendment, males were also scooped up into the penal system for consensual homosexual acts, too; and this makes it difficult, when looking at the statistical record, to disentangle genuinely abusive behaviour from sexual acts to which both partners consented. It is thus hard to interpret historical trends in arrests for gay sex, and any consequent certifications under the Mental Deficiency Act.

The years in which John S— was active seem to have represented something of a high point in arrests and prosecutions of gay men. Perhaps it was this that prompted Mrs S— to seek his certification and guardianship. A government inquiry in 1925 stated that nationally, prosecutions of 'unnatural offences' had doubled between 1909 and 1925. In London, the number of males prosecuted for importuning other men for sex stood at around eighty each year from 1922 to 1927 inclusive, but then dropped to fifteen in 1928 and ten in 1929. This fall was the result of Metropolitan Police officers coming under increased criticism for the unfair surveillance of civilians, entrapment and even allegations of lying in court; staking out, arresting and making a case against active homosexuals no longer seemed to be worth it for the average constable, and so prosecutions in London fell.

'Gross indecency', or 'unnatural offences', do not predominate in the available records of males certified under Section 8. For example, a dataset from the London Metropolitan Archives breaks down the offences in 233 magistrates' court cases in 1923 in which young males were deemed to be mentally defective thus: stealing, 102; 'wandering and begging', sixty-one; indecent exposure, eighteen; indecent assault, seventeen; gross indecency/unnatural offences (i.e. with other males), thirteen; assault, twelve; drunk and disorderly, ten. (The average mental age of all the lads examined was put at 8.4. For girls in the same dataset, it was 8.7.)

The surviving case histories of pre-adolescent boys put into colonies as moral imbeciles sometimes record that one factor in their uncontrollability was their sexualised behaviour towards other children – but this tended to be sexual aggression towards little girls more often than towards other little boys.

Crime statistics for the years 1931 to 1936 inclusive (compiled for the purpose of finding out how many people convicted were found to be mentally defective during or after the trial) reveal that the offences being committed by those labelled defective were overwhelmingly property crimes. Burglary, theft, robbery and fraud account for forty-two per cent of crimes committed by the allegedly defective. The second-largest crime category committed by this group were sexual offences – rape, sexual assault, carnal knowledge of a child, 'unnatural' (homosexual) acts, bestiality, incest, indecent exposure and indecent behaviour; these comprised twenty-five per cent of all crimes by 'defectives' (and were almost always committed by males). Vagrancy, rough sleeping and begging made up the other sizeable category, at eight per cent.

Some eighty-four per cent of defective people convicted of crimes were male, in line with the male non-defective convicted percentage of eighty-seven. All types of offence committed by those found to be defective were, by a massive margin, committed between the ages of fourteen and twenty-eight; over half were committed between ages sixteen and twenty-one.

Crime statistics are notoriously problematic. Obviously, only offences that are reported can make it into official datasets: it is likely that most crime passed (and continues to pass) unreported. So, an attempt to make a meaningful comparison between acts committed by the allegedly defective and the 'normal' population (the term used by government at this time) between 1913 and the mid-1930s is hedged about with risk. For one thing, as we have seen, subjectivity and arbitrariness played a significant role in deciding who could be labelled mentally defective. In around eighty per cent of cases, a defendant had never previously been suspected of feeble-mindedness until being brought up on a criminal charge – it was only the illegal act that had prompted the question of their mental state. With the criminal act itself being proposed as a symptom of mental defectiveness, wasn't there a danger of tautology? Because you have committed a criminal act you are mentally defective, and because you are mentally defective you have committed a criminal act.

We must bear these caveats in mind as we read the Board of Control's findings for the early 1930s. Most strikingly, in 1931, of 589,657 people found guilty of criminal offences, just 897 were mentally defective. This was cause for some self-congratulation by the Board, as it suggested that early certification and institutionalisation of 'feeble-minded' children had indeed prevented them committing criminal acts later in life – that society had successfully been protected from the undesirables. Yet, between the ages of sixteen and twenty-one, defectives were twice as likely to offend as 'normal' people in that age cohort; and in crimes of violence, including sexual violence, defectives in every age group were seven times more likely to offend. However, outside of that stage of life, and after discounting property crimes, violent and sexual offences, and vagrancy/begging, 'normal' people were responsible for significantly more crime and anti-social behaviour than those deemed defective – by a factor of fifteen to one.

The figure of the perverted and violent 'mental case' came to dominate media coverage. From the late 1920s onwards, as newspapers became increasingly explicit in their reporting of sexual violence, particularly sex killings of little girls, the depraved simpleton came to centre stage.

13

'AND THEY CATCH 'IM, AND THEY SAY 'E'S MENTAL': THE SEX OFFENDERS

'Child Killers at Large' ran the headline in the *Daily Herald*: '5,240 certified mental defectives are free. Parents are asking "why?"' Tall, with a lumbering gait and huge warty hands, John Straffen embodied the nightmare figure of the lethal, amoral child-in-a-man's-body – reminiscent of Bela Lugosi's incarnation of Frankenstein's monster. Like the monster, Straffen ended the life of a little girl; and then another, and then – after escaping from Broadmoor Hospital – still another. His crimes reanimated the 150-year-old conundrum about culpability and mental functioning. During Straffen's trial, in 1951, the public and the newspaper press feasted on the missed opportunities to deal effectively with his destructive impulses towards girls, and on the medical experts' seeming inability to agree on his treatment, both before and after the murders he committed.

Born the third of six children in March 1930, John Straffen had been slow to learn to talk, was declared 'backward' at school and had thick, mumbling speech. At age eight he was referred to the city of Bath's child guidance clinic because of his pilfering and truanting, and in June 1939 he came before the juvenile court, where he was sentenced to two years' probation for stealing a little girl's purse; his probation officer, with many years' experience, stated that Straffen was unlike any of his previous cases – he was 'a real problem', he said. A psychiatrist agreed and recommended he be certified as a mental defective because of his low IQ and apparent failure to understand right and wrong. The family background was deemed unsuitable for Straffen to remain at home under licence or supervision: his father (a soldier) and mother were burdened with the care of six children (one of their daughters was a 'high-grade mental defective'), in overcrowded conditions.

Tested in 1940, ten-year-old Straffen was declared to have a mental age of six. Throughout his adolescence his manner was described as cheerful and friendly, with no 'harmful characteristics'. At Besford Court Catholic Mental Welfare Hospital staff described him as timid, docile and friendly, but solitary and prone to extreme sulkiness when scolded. (The Straffens were Protestants, but national lack of bed availability meant that Besford was the best available option.) Peter Whitehead, who was at Besford at the same time, would later recall Straffen as 'a tousle-haired boy with thin features, who was always reserved, making no friends and rarely joining in games'. Peter tried to talk to him, but he rarely responded, and so the other boys 'learned to ignore him, as if he were not there'.

At age sixteen John's mental age was nine-and-a-half, and it was felt that family life might be good for him so he returned home to Bath in March 1946, obtaining a job as a machinist at a clothing manufacturer. Ten months later he was off the rails again, breaking into empty houses and stealing – not for gain, just for ownership of the items. On 27 July 1947 he put his hand over the mouth of the thirteen-year-old girl he had been playing with and said, 'What would you do if I killed you? I have done it before.' Six weeks later, Straffen quarrelled with another young girl and in retaliation strangled five of her father's chickens. Upon arrest, and without prompting, he boasted to the police of the thefts he had perpetrated – no one had known about them previously. He also claimed that he had committed a number of sexual offences, but the psychiatrists dismissed this. The police did look into the matter, but could find no substantiation of his statement.

The local authority sent him to Hortham Colony near Bristol. Very oddly, his certificate stated that he was 'not of violent or dangerous propensities', even though violence towards a girl and the killing of livestock was known of. In August 1950 he absconded and returned to the family home. When the police arrived to take him back he resisted violently. But six months later he was licensed to the care of his mother. She asked the Board of Control to release him from his certification, but when they came to his home to re-examine him, they found that his mental age was just ten. The

Board told his mother that they would look at the case again in six months' time.

Five days later, on Sunday 15 July 1951, Straffen set off on foot for the local picture house, where Douglas Sirk's *Shockproof* was playing. He walked past a meadow where five-year-old Brenda Goddard was picking flowers. He set upon her, began to strangle her and 'because she did not scream' (he later said) he beat her head against a stone wall. Without attempting to hide her body, he went and enjoyed *Shockproof* and then went home.

Three weeks later, at a screening of a Tarzan film, he lured nine-year-old Cicely Batstone away from the cinema and onto a bus, alighting at a meadow, where he laid her down and strangled her from behind. He then headed off to get some fish and chips.

Cicely's body was found at dawn, and Straffen was arrested the same morning. His childish lies to the police about how Cicely had importuned him away from the cinema got him nowhere and – without anyone raising the topic of Brenda Goddard – he boasted of how he had attacked Brenda.

He was charged with the two murders on 17 October 1951 but was found unfit to plead because of his mental deficiency. It would have been like trying a child, the doctors decided. He was sent to Broadmoor Hospital to serve time at His Majesty's pleasure. Six months later he made his escape. Reaching the village of Farley Hill, seven miles from Broadmoor, a few hours later, he spotted five-year-old Linda Bowyer playing on her bicycle, and strangled her.

Berkshire County Council had been expressing concern about the potential danger to the public of recent changes at Broadmoor: there was no warning alarm system for when an inmate escaped. The introduction of civilian clothing at the hospital meant that any escapee would be much harder to spot. A subsequent enquiry found that some of the locks within the hospital were obsolete, and that there was one rather obvious escape route over the wall – the very one that Straffen, with a mental age of ten, had figured out. (That said, there were only twelve escapes between 1900 and 1954.)

The slackness of the Broadmoor regime provoked public fury; and there was contempt, too, for the legal and psychiatric systems that permitted Straffen to avoid standing trial for murder in October 1951, but being declared fit to plead and be tried just eight months later, in July 1952. The public in the gallery at his trial, at Winchester Castle, periodically made clear their disdain for expert opinion.

Straffen presented a conundrum. The Mental Deficiency Act, as we have seen, permitted a defendant to be sent to an institution instead of prison upon conviction, or to avoid a trial by reason of being unfit to plead because of feeble-mindedness; but Straffen was being tried on a capital charge – and there was no precedent for a mental defective being put on trial for murder. The only option for his defence counsel was to attempt a very broad use of the M'Naghten Rules; these were formulated in the 1840s to establish a defence on the grounds of insanity. With the M'Naghten Rules, the defence had to demonstrate that at the time the act was committed, the accused's mind was so diseased that s/he could not know what s/he was doing, or that s/he did not know that the act was wrong. M'Naghten concerned mental illness, not mental defect; and as lawyer and doctor Letitia Fairfield (author of the Notable British Trials account of the Straffen case) stated, in the early 1950s, far too many in the British legal system (let alone the jury, the press and the public) were still unable to differentiate between mental illness and mental defect.

Fairfield wrote that if a defective with a very low IQ were to commit a murder, there was no hesitation in regarding him as unfit to plead; both Broadmoor and Rampton held inmates of this type, she said. However, if he had a mental age of seven or older, it was extremely difficult to maintain he didn't know that what he was doing was against the law. Straffen, with an established mental age of ten, had boasted in his police interviews that he had committed the killings in order to annoy the police – whom he hated. His acts had shown cunning and deception, and subsequently he had successfully planned and executed an escape from Broadmoor. 'Mental defectives are by definition NOT insane', wrote Fairfield.

All the doctors, whether called by the defence or the prosecution, agreed that Straffen was certifiable under the Mental Deficiency Act as feeble-minded; however, they said that he knew that he was killing the girls, and furthermore, he knew that this was wrong.

Letitia Fairfield believed there was 'no decisive test which would place the high-grade defective once and for all outside the category of normal people'. For what it's worth, Fairfield had no time for the 'psychopath' category, in which the majority of psychiatrists believed, by the early 1950s: 'No one has yet been clever enough to define a psychopath', she wrote. She came pretty close herself, though, diagnosing a huge resentment within Straffen which she said he knew how to keep hidden behind a façade. This resentment, she believed, only found expression 'in vindictive violence directed against some creature who cannot retaliate, such as an animal or a child'.

Because the M'Naghten Rules had not been updated or expanded, the instructions from the judge to the jury during the Straffen trial still used the word 'insane': at the time he committed the killings, was he insane within the meaning of the criminal law?, the judge asked them. The ten men and two women jurists took twenty-nine minutes to decide he was guilty. The judge put on his black cap and sentenced Straffen to death.

Britain had not executed a certified mental defective before. Timothy Evans doesn't count, since, despite having a mental age of ten-and-a-half, he had never even been suggested for examination and certification under the Mental Deficiency Act. Evans went to the gallows two years before the Straffen trial, for the murder of his wife, Beryl, and baby daughter, Geraldine – no one yet suspecting his landlord, Reg Christie, who had turned 10 Rillington Place into an informal burial vault.

Straffen's execution date was set for 6 September 1952, and nine days before – courting mass public anger – Tory Home Secretary David Maxwell Fyfe issued a reprieve. Letitia Fairfield believed that Fyfe had been swayed by the testimony of Dr Alexander Leitch, who had pointed out that most nine-and-a-half-year-old boys would have far larger capacity than Straffen to learn from past mistakes and to foresee the results of their actions. He was qualitatively different to 'normal' nine-year-olds. Straffen's extraordinary conduct during and after the crimes 'was fatuous in the extreme' and indicative of a lack of appreciation of the *meaning* of his killings. (Famously, Fyfe the following year would refuse to reprieve Derek Bentley, a young man with a mental age of ten years and four months who was present when his friend shot dead a policeman. 'Let him have it, Chris', Bentley had said, almost certainly meaning, 'Chris, hand over the weapon as requested.')

Straffen went on to be Britain's longest-serving prisoner, dying in 2007 after fifty-five years and three months in gaol (a record since broken).

Although no evidence of sexual assault was found on the three bodies, Straffen's victim typology was always small girls and, as Fairfield noted, in the act of strangulation 'he derived from it some emotional gratification, probably of a specifically sexual nature. Some very good judges of human nature who have been in contact with him think this is the predominant motive in his crimes.' Straffen's murders were sex killings, and the public surmised as much. His reprieve added to a growing popular view that vicious killers and their duped supporters were using psychology and psychiatry to make sure natural justice was evaded. 'And they catch 'im, and they say 'e's mental', is how one of the teenage South London lads in Karel Reisz's 1959 documentary film *We Are the Lambeth Boys* contemptuously describes this notion. (Conversely, the executions of Evans and Bentley added to concern that the judiciary and psychiatric 'expert opinion' were making arbitrary decisions.)

And then it happened all over again. Zalig Lenchitsky had been certified mentally defective and placed in an institution in 1938 after being caught in the act of assaulting a little girl – she had screamed very loudly, which alerted the neighbours, who rescued her. After seven years in Farmfield and then Darenth Park colonies, Lenchitsky was discharged to the care of his father in East London, and to a manual job, as a factory worker. On 19 September 1953, now aged forty-two, but with a mental age of nine, he lured five-year-old Wendy Ridgewell to his home at 71 Teesdale Street, Bethnal Green, removed her clothes, tied her up and suffocated her. He placed her body in a box and left it in the shed in the back yard.

As with the Straffen murders, there was no forensic evidence of sexual assault, but the crime was presumed to be sexual in origin. 'Something came over me in my head', Lenchitsky later told the police.

Perhaps aware of the ongoing public anger at the Straffen case, there was no attempt to prevent Lenchitsky's case going to full trial; it was heard at the Old Bailey. The Brixton Prison medical officer who found him fit to plead stated that 'while it is the case that this prisoner is feeble-minded and in consequence his reasoning powers, his judgment and his self-control are sub-normal, he is capable of knowing the nature and quality of his acts and whether they are right or wrong. In particular, he has a knowledge, though impaired, of what acts are against the law . . . He cannot, or will not, tell me why he attacked the victim.' His defence team argued that with his low IQ, Lenchitsky did not intend Wendy's death to follow on as a consequence of gagging her; he was incapable of planning such an outcome. But the jury found him guilty of murder, recommending mercy (perhaps with Evans and Bentley in mind). They believed that he had been capable of forming an intention to commit harm; but they did not want him to swing for it. The Home Secretary (Maxwell Fyfe again) agreed, and his death sentence was commuted in January 1954. Lenchitsky died at the age of seventy-seven, in 1989.

The 1957 Homicide Act introduced the plea of diminished responsibility, which brought mental deficiency more in line with the rules on mental illness.

***

Reader, you will have noticed that these horrific cases make a strong argument against the main thrust of this book – that certification and institutionalisation under the Mental Deficiency Act was misused, and kept many individuals under lock and key when they should have been out and about in the community, living the fullest life possible. I think that the administrative machinery for dealing with people who had shown propensities for serious, violent, crimes could in fact have provided a civilised solution, so long as it was wisely and consistently administered – and properly funded. Supervision of both Straffen and Lenchitsky when they were released to their families, on licence, was woefully inadequate. And there is a good argument that, given both men had assaulted small children before they went on to kill, any form of licence – ever – was unwise. Detention into old age for such men – in humane and well-run institutions – seems like a reasonable trade-off for the safety of the young, or otherwise vulnerable, I am almost persuaded.

***

The lack of accommodation in mental deficiency institutions had been a problem from the moment the Act had come into operation, on 1 April 1914. A costly war immediately diverted money and capacity, as already noted; and the 'Geddes Axe' of 1922 saw one of modern British history's largest squeezes on government expenditure. The Second World War had, as with the First, drained money and attention from mental deficiency services. The national shortfall in mental deficiency beds reached 9,000 by early 1954. The fear, whipped up by an increasingly sensation-hungry popular press, was that, *á la* Straffen and Lenchitsky, the women and children of Britain were ever more vulnerable to the depredations of the prematurely released, sexually out-of-control man-child. It's all the more maddening to recall that hundreds, perhaps thousands, of mental deficiency beds were at this time occupied by needlessly detained women who had had children out of wedlock. I am quite unable to fathom why the official mind did not make *their* release a priority – I've seen nothing in any paperwork to suggest why this solution was not considered. For those who had no family to return to, sheltered housing or staffed hostels could have been pioneered to a far greater extent than they were.

Even after the creation of the NHS in 1948, mental deficiency remained a neglected public service. In the late 1950s, magistrate and Conservative MP for Tynemouth Dame Irene Ward repeatedly harangued her own party's Ministry of Health and Home Office on the topic, her especial concern being the lack of accommodation in the north-east of England for low-IQ boys and young men whose aggressive sexualised behaviour was a threat to others. Mrs E— J— of Whitley Bay, a widow with adolescent twin sons who had both been certified as mentally defective, had written to Ward in despair because of the local authority's inability to find institutional places for them. The behaviour of David and Charles J— had led to many complaints by locals, and David had already been held on remand because of his propensity for 'molesting females', including his thirteen-year-old sister.

The Ministry sympathised with Ward, and acknowledged that the north-east had been neglected in these matters. When the NHS had been founded, the Newcastle region had had 0.46 mental deficiency beds per 1,000 head of population, in comparison to the national figure of 1.1 per thousand. The Ministry mollified Ward by pointing out that a large building programme was now under way in her locality, and that there were thirty-one per cent more patients in mental deficiency institutions in the Newcastle area in 1958 than there had been in 1948; over the country as a whole the increase was only about twenty per cent.

Next, she tried Rab Butler, Tory Home Secretary, telling him that one of the recent fiscally forced decisions by Tyneside magistrates had led to the sexual murder of a young boy by a youth who was on probation awaiting a place in a mental deficiency colony. She told Butler that magistrates were

> often forced into decisions which they themselves deplore because the Treasury in its isolation makes it impossible for the necessary accommodation to be made available to prevent these ghastly things happening, and I think it is time that somebody said so in no uncertain terms... One of my constituents was murdered because a young man of abnormal sexual habits was put on probation. I'm very annoyed with the judge and I think HM government, whose fault it is, should apologise to the magistrates.

Her prompting on behalf of Mrs J— appeared to work, partially, as David J— was found a place at Prudhoe Hall. But Ward remained concerned that no similar accommodation had yet been found for his twin:

> I do not see why a young girl's life should be worried, or the parents either, by the unfortunate situation of the boys. Nor do I see why any risk should be run by this child or any young married women living in the vicinity who have already had advances made to them by these boys... We cannot get mentally weak, sexually perverted young boys who have been set to go to a proper home into one, as there is no room.

Ward had a reasonably loud voice within governmental circles. She would be the longest-serving British female MP until Gwyneth Dunwoody broke her record in 2007; and Ward was awarded the OBE in 1920 and DBE in 1955. So she could write informal, hectoring letters addressing the Home Secretary simply as 'Rab'; and be confident in her calls for the government to take far greater notice of the safety of potential victims of sexual attacks as part of her campaign for increased mental deficiency accommodation. However, there was in some quarters of government an already established acknowledgement of the impact of sexual crime on its victims, dating to at least 1925. In that year, the Departmental Committee on Sexual Offences Against Young Persons issued its report, having been convened to consider the inadequate sentencing for those found guilty of sexual crimes, as well as the difficulties of prosecuting such cases: the acquittal rate was far higher for sex crimes than for any other type of offending, for all the reasons that are so depressingly familiar in the UK today.

Fully aware that sexual crimes were likely to be drastically under-reported, and allowing for the colossal upheavals of the 1914–18 War, it was the considered opinion of the Departmental Committee that since circa 1909, the huge increase in sexual assault convictions disguised the true prevalence of rape: in order to try to secure any kind of conviction, rape was too often downgraded to the lesser charge, and in this way, 'The gravity of a sexual offence is masked.' Serious sexual assaults against boys and girls were also reduced to lesser charges, usually to permit the child to avoid the anxiety, delay and publicity that an assize court appearance would cause; summary justice before a magistrate was faster and comparatively under the radar.*

As mentioned in chapter 9, a shift was starting to take place in appreciating the psychological impacts of child sexual abuse – observations that were different in substance to the late-Victorian and Edwardian notion of the 'corruption' of children by early exposure to sexual matters (although that view was also still being expressed in the 1920s). The report of 1925 found that

> in many children, the commission of the offence causes a shock to the nervous system varying according to age and nature of the child, and we have been told that, with some, chorea [involuntary muscle movements], sleeplessness and other nervous disorders have resulted. Even if no physical injury is apparent, there may be grave mental disturbance. We have evidence of cases in which children who have been subjected to more than one indecent assault have had their moral sense impaired and for this reason become a source of danger to other children.

Quoting evidence from the Medical Women's Federation, the Report of the Departmental Committee on Sexual Offences Against Young Persons continued:

---

* Until 1922 incest trials were held *in camera*, following the passing of the 1908 Incest Act. However, it became apparent that this meant that the fact incest was illegal (and not just unethical or anti-social) was not getting through to the general public. This is why the press were allowed in to court to report, with names redacted to protect the victim's identity.

> To awaken or excite the dormant passions of a child or young person... prematurely is to confound its sense of right and wrong, and to give it a false view of human and social relationships and duties which it is very difficult to correct later. The physical and mental balance is often upset, and healthy development hindered. Not only unfortunate impressions but severe neurosis may persist in later life as a consequence of such experience.

The report additionally noted the impact of both physical injury during sexual assault and of pregnancy and childbirth on an immature body. Meanwhile, Ministry of Health data on venereal disease in children showed that while the bulk of this was congenital syphilis, gonorrhoea, as an acquired infection, was a true indicator of a sexual assault upon a child. The Ministry of Health statistics showed that girls between babyhood and fourteen years of age were infected with gonorrhoea fifteen times more often than boys. The agencies reporting on such matters were of the opinion that 'the superstition that connection with a virgin will cure a man' of venereal disease was thankfully dying out.

The Committee had interviewed many workers in the field, and most had said that the common belief that the men who committed sexual offences were either insane or mentally defective was incorrect. The majority of convicted sex offenders had no 'abnormality' that could be ascertained. Of 108 men remanded to prison in the three years ending March 1924 for indecent assault, incest or carnal knowledge of a girl or boy under age sixteen, none was found to be insane and only eight were considered certifiable as mentally defective. The cases of indecent exposure showed a somewhat larger proportion of mental trouble: of 150 men remanded to prison for this offence, fourteen were declared insane and twenty-four would be certified as mentally defective.

The Committee was concerned that focusing on the mentally defective defendant shifted the emphasis away from the more concerning truth that all types of men committed sex offences.

> We have had such varied types of offenders brought to our notice that we consider that there can be no one method suitable for dealing with all cases...
> Some offenders have been mentally unstable or weak-minded, whilst others

have been well educated and intelligent. Loafers and vagrants have been concerned in some sexual offences, while professional or business men of apparent respectability have been found guilty of the same offence. There have been difficult cases of young boy offenders and also cases of elderly men.

However, tallying with the findings of the 1930s crime survey (noted in chapter 12), of the male prisoners in the 1925 sample who were discovered to be mentally defective, one-quarter had been convicted of a sexual offence. 'Other expert evidence confirms these statistics in showing that among the insane and mentally defective, sex obsession or sex perversion is somewhat common, and that in cases of indecent exposure there is a fairly large proportion of men suffering from mental disease or defect.'

CASE A.		CASE B.		CASE C.	
2.11.08	Indecent exposure, 21 days.	14.9.06	Indecent exposure, 14 days in workhouse.	18.6.88	Indecent exposure, 3 months.
19.8.09	Indecent exposure. (2 cases) 3 months and 14 days (consecutively).	17.9.10	Indecent exposure, 3 months' imprisonment.	4.2.89	Indecent exposure, 3 months.
17.10.13	Indecent exposure, 9 months.	3.8.11	Indecent exposure, 3 months' imprisonment.	2.1.93	Assault on girl age 9. 2 months.
5.9.14	Indecent exposure, 6 weeks.	31.10.11	Indecent exposure, 14 days' imprisonment.	16.4.95	Indecent exposure, 6 months.
12.1.15	Indecent exposure, 9 months.	10.7.12	Indecent exposure, 14 days' imprisonment.	14.7.02	Indecent exposure, 9 months.
27.10.15	Indecent exposure, 12 months.	9.8.12	Indecent exposure, 1 month or fine.	10.7.03	Indecent exposure, 6 months and 9 strokes with birch.
10.10.16	Indecent exposure, 6 months and 15 strokes with cat.	17.10.12	Indecent exposure, 1 month or fine.	3.1.06	Indecent exposure, 12 months and 12 strokes with birch.
12.10.21	Indecent exposure, 3 and 3 months (consecutively).	17.7.13	Indecent exposure, 3 months.	4.1.07	Indecent exposure, 12 months.
5.11.23	Indecent exposure, 3 and 3 months (consecutively).	30.8.16	Indecent exposure, 6 months and strokes with cat.	20.2.08	Indecent exposure, 11 months.
28.5.24	Indecent exposure, 3 months.	7.7.20	Indecent exposure, 12 months.	20.5.13	Indecent exposure, 3 months.
20.8.24	Indecent exposure, certified insane.	9.1.22	Indecent exposure, 12 months.	9.10.16	Indecent exposure, 6 months.
		6.4.23	Indecent exposure, 12 months.	20.3.19	Indecent exposure 6 months.
		14.4.24	Indecent exposure, 12 months.	19.6.20	Indecent assault on 2 girls, 6 and 6 months (consecutively).
				10.6.21	Indecent exposure, 3 months.
				6.1.22	Indecent exposure, 12 months.
				16.11.23	Gross indecency with male person. Order under Mental Deficiency

The 1925 Departmental Committee on Sexual Offences Against Young Persons published data revealing long histories of repeat indecent exposure, indecent assault and gross indecency 'across very many years, and no punishment appears to have acted as a deterrent'.

The 1925 Committee unearthed data concerning men with very long histories of repeat indecent exposure, indecent assault and gross indecency (see the table on p. 178), their offending taking place across 'very many years, and no punishment appears to have acted as a deterrent'. The Committee sought a great expansion in the amount of mental testing for those on trial for these crimes, and 'prolonged detention of men who appear quite incapable of abstaining from indecent exposure or from committing repeated indecent assaults on children'.

Aware that it was quite un-English to implement indeterminate sentencing, they nevertheless came to the conclusion that 'a period of prolonged detention in a special institution might occasionally effect a cure. In any case, it would protect the public more effectively.' They did not seek to remove the court's sentencing discretion, they said, but short sentences for crimes against women and children were outraging the public, especially in comparison to property crime sentences: 'It sets a wrong moral tone for society to value property more than the victims.'

The Committee members, like many who worked in the administration of justice, had strong reservations about the 'moral imbecile' category – not accepting that an average or high IQ could permit someone to be described as any kind of imbecile. Surely they were simply criminal – like any other wrongdoer, only perhaps more insidious.

> High-grade feeble-minded persons are exceedingly difficult to detect as such by those unaccustomed to dealing with them, and magistrates are apt to consider that if a man or woman has not an imbecile appearance and can answer ordinary questions, he or she cannot be a person requiring 'care, supervision and control for their own protection and for the protection of others'. This also applies to that very difficult class, the moral imbecile, whose intellectual attainments are on a much higher plane . . . Our recommendation for the appointment of skilled mental advisers should help to overcome these difficulties of diagnosis.

That turned out to be a vain hope.

Sex-offending defectives should not be offered guardianship or be allowed out on licence, they said, no matter the shortage of institutional places. Another unrealised aspiration.

In the late 1920s, another solution bubbled to the surface once again – to deal with sex offenders and in fact the whole range of undesirables: sterilisation.

PART 4

SOLUTIONS

14

THE GATESHEAD CASTRATIONS, OR PAGANISM AND THE KNIFE

An extraordinary number of people who ought to have known better assumed that to sterilise a man removed his sexual urges. Churchill had thought this back in 1911 when he was Home Secretary and had interested himself in the case of Alfred Oxtoby. Oxtoby, classified as a 'low-grade' imbecile, had been convicted of attempted bestiality with a mare at his blacksmith father's smithy and was being detained at His Majesty's pleasure at Broadmoor. 'He is a typical example of the village fool with dangerous sexual tendencies', stated his case file. It had to be explained to Churchill that performing a vasectomy upon Oxtoby would not lead to loss of libido; the Home Office advised him that 'certainly the procreative power would cease, but its effect in lessening the sexual appetite is extremely problematical – probably it would have no effect at all. The patient, being non-*compos mentis*, could not give his consent . . . Even if this remedy received public approval in such cases as the present, it could only be carried out in a fully equipped hospital and by a competent surgeon.'

Almost twenty years later, 'a certain learned judge', according to the Ministry of Health, was seizing every opportunity to state from the Bench that the sex offenders who came before him would, for the most part, not have committed such acts if they had been sterilised – a level of biological ignorance in a highly educated man that astonished the Ministry. This was not 'a problem that can be solved by a surgeon', the Ministry stated.

The judge in question was Justice McCardie. While sentencing at Aylesbury Assizes in July 1929, McCardie pronounced: 'Sterilisation would help greatly towards reducing the grave state of mental defectiveness which exists in many parts of the country. There is nothing more horrible in our state of civilisation than that men and women should be able to beget eight or nine or a dozen illegitimate children with this horrible taint.' In the cases before him, Annie Hall, forty-one, of Aylesbury, who had eight illegitimate children, had just killed the ninth; but the medical officer of Holloway Prison had refused to certify her feeble-minded, on the grounds that he knew 'nothing of her history'. That is to say, he had no eugenic evidence – her mental tests and her crime were clearly not sufficient for him to make such a diagnosis.

It is tempting to view this refusal as the medical officer being deliberately obstructive precisely in order to show how unworkable eugenics could be 'on the ground': it was impossible to discover a familial 'pedigree' in the type of mobile and rootless population from which Hall hailed. Many MOs, and many GPs, had never accepted eugenic arguments, and their everyday interactions with 'undesirables' confirmed them in their scepticism. Nevertheless, Judge McCardie continued that he hoped 'that the scope of the Mental Deficiency Acts would be greatly enlarged to mitigate this growing peril of the unintelligent, the inferior and the almost insane'.

At around this time, castration to eradicate sexual impulses was undertaken illegally at the High Teams Poor Law Institution at Gateshead, north-east England. William W—, twenty-two, a certified 'imbecile'; Richard P—, fourteen, on remand for an indecent assault; and eight-year-old Henry L—, an 'imbecile' with epilepsy and unable to speak, were castrated in the early summer of 1930. Each was described as highly sexually aggressive, and all three sets of parents had requested the operation in the belief that it would improve their sons' behaviour. (Only William W— had actually been certified under the Mental Deficiency Act.) Parental consent – any consent – was no defence in law, however, as the removal of testes was only permissible if there was a serious risk to physical health if the operation was not performed.

The Ministry of Health investigated, and reprimanded the doctor concerned; when he agreed not to repeat the procedure, the matter was dropped. Whitehall warned the authorities at the Gateshead institution that they were 'running a very great risk' if these operations should come to the attention of 'the anti-sterilisation organisations' and the 'liberty of the subject men' (like Josiah Wedgwood – still fighting the good fight from the back benches. By now, good-humoured contempt of this kind was often showered upon those who spoke up to support the liberty of the subject).

Something that buoyed the various *supporters* of sterilisation were the requests from parents of learning-disabled children for operations to be performed upon them so that they could not become pregnant or cause a pregnancy; or in the mistaken belief that sterilisation could prevent them committing sexual assaults. Sterilisation proponents claimed that these parental and familial requests showed that the British public felt a much greater approval of the operation than government assumed. The Board of Control received correspondence from various local authorities seeking advice on how to respond to such applications from family members. From Peterborough came a medical officer's letter asking whether Irene W—, described as 'a Mongolian Imbecile', aged ten years and eight months, could have her fertility ended:

> She has now begun to menstruate, and her parents wish her to be sterilised. May I ask whether operation or [Roentgen] X-ray is recommended by the Board of Control? Is there any place in London, or the Provinces, where a large number of cases have been done? I should be glad to have the views of the Board of Control on the matter. I should say that the question was asked quite voluntarily without any suggestion on my part.

The widowed mother of twenty-nine-year-old William D—, who had been committed to Rampton in 1926, wrote to the Board of Control asking that he be sterilised so that he could be safely released. 'He worked very well for me when he was at home and gave me all he earned . . . Is it not time for him to be released and to give him a chance in life?' The Board replied, as it replied to all similar requests, that no such option was available and that an operation of this kind would at that time (the summer of 1931) be illegal.

These parents did not see sterilisation as an infringement of personal liberty: to free their offspring, or relatives, from having unwanted babies seemed to them to be a positive – just as Winston Churchill had claimed. In many cases, these families believed that a sterilised individual would be permitted to live safely out in the community, unable to procreate and thus free to enjoy a romantic and even a sexual life. 'Sterilisation ought to be regarded as a right and not as a punishment', is exactly the finding of the Brock Committee, which began its inquiry in late 1932 (under the chairmanship of Sir Laurence Brock, head of the Board of Control) into the causes of feeble-mindedness and the possibility of legalising voluntary sterilisation. The Committee reported in 1934. To the delight of the Eugenics Society, the Brock Committee recommended the legalisation of voluntary sterilisation for so-called 'high-grade' mental defectives – an operation that could only take place with the permission of the person to be sterilised. There was no need to perform the operation upon the low-grade, the Brock Committee Report stated, since their lives were spent in sexually segregated institutions and so they were highly unlikely to produce offspring.

It was for social reasons that Brock wished voluntary eugenic sterilisation to be legalised. Mental defectives made terrible parents, the Report said, and while it was still extremely difficult 'to say with certainty that the genetic endowment of any individual is such that it must produce a given result', they felt certain that 'the vast majority of defectives are temperamentally and socially unfitted for parenthood . . . Mentally defective and mentally disordered parents are, as a class, unable to discharge their social and economic liabilities or create an environment favourable to the upbringing of children.'

The Report refuted the argument that voluntary sterilisation could lead to the mass release and freeing up of beds in institutions. Most 'low-grade' cases could never live independently; and the 'high grades' and 'moral defectives' who would be most suited to the operation would not be psychologically stabilised by being neutered: 'The unstable and antisocial defective remains unstable and antisocial. The thief remains a thief. The erotic girl or youth will still need institutional care. The impossibility of procreation will not save them from being a social menace . . . Their antisocial tendencies would be more effectively controlled by segregation.'

The Brock Committee came out firmly against any compulsion in sterilisation; for one thing, it would cement the idea that the mental deficiency authorities were to be feared, and this might have the effect of 'driving defect underground', with parents refusing to permit their children even to be ascertained, let alone put into a home.

But how could 'informed consent' by the patient apply, when the very nature of feeble-mindedness meant that reasoning powers were impaired? The Report of the Brock Committee descended into tendentious arguments on this point: 'So long as there is no unfair pressure and no patient is forced or bribed to consent, it seems to us mere casuistry to discuss how far the patient fully appreciates all the implications of consent ... We are anxious that no pressure should be brought to bear on the patient and we are convinced that anything in the nature of veiled coercion will do nothing but harm.' But how likely was it that anyone would agree to be made infertile if they fully understood its implications? And if they didn't understand the implications, the procedure was unethical in the first place.

The Report added that if the doctor believed that the feeble-minded patient really hadn't understood the issue, the parent or guardian should have the power to consent for them. This makes the word 'consent' do a great deal of heavy lifting.

Anticipating the argument that only the poor would be targeted for voluntary sterilisation, the Report pointed out that the well-off had been able to pay to control their fertility, including sterilisation if they so wished, while the poor could afford no such medical intervention to limit their family size: why shouldn't 'the poor man, a victim of an inherited physical or mental disorder', not be permitted to opt for the operation – free of charge? Vasectomy was touted as comparatively straightforward; for women, though, the four types of sterilising operation had greater potential for physical and psychological after-effects – cutting of the Fallopian tubes (salpingectomy), removal of the ovaries (ovariotomy, or oophorectomy), removal of the uterus (hysterectomy), and radiation by Roentgen rays (X-rays; this method did not become common, since the dangers to health became apparent very early on).

Eugenicists would spend the rest of the 1930s pushing for parliamentary action to legalise voluntary sterilisation; many of them hoped this would open the door for compulsory measures. One of the ultimate reasons that they failed to achieve their aims was the widespread suspicion that voluntary measures would indeed facilitate legislation to permit involuntary sterilisation. It was vital that the eugenicists should not achieve even the most modest of their aims. And from the start, they were up against formidable opponents. As with the debates of 1911–13 on the Mental Deficiency Bills, the Roman Catholic Church was the most implacable and vocal opponent of sterilisation. The papal encyclical *Casti Connubii*, emitted on 31 December 1930, reiterated the Vatican's opposition to eugenics and to all artificial attempts to control fertility. The British Medical Association opposed even voluntary sterilisation (with a few dissenters). The *British Medical Journal* was hostile. (The Royal College of Surgeons and the Royal College of Physicians were, though, broadly in favour.) And this time around, the organised left – the Labour Party and the trades unions – almost unanimously came out in strong opposition to renewed attempts to curtail permanently the fertility of any section of the population. They now fully appreciated the class aspect of the sterilisation proposals; the Cassandra figure of Josiah Wedgwood was vindicated by most of the left by the early 1930s.

So where had the left's appreciation of the allegedly 'progressive' nature of eugenics gone? What had happened in the intervening years that would make a speech like that of Will Crooks MP in 1912 (mentioned in chapter 7) now almost impossible for a working-class Labour member to make in parliament?

For one thing, the evidence that would gift certainty to eugenical thinking had not arrived. If anything, it was the complexity of factors behind alleged 'feeble-mindedness' that had begun to take hold. The more that was discovered about nutrition, housing conditions, exhaustion, anxiety from overwork, lack of educational opportunities and the impact of physical diseases, the more complicated their interactions with heredity became. Serious pushback by scientists including J.B.S. Haldane, Julian Huxley and Lancelot Hogben was, from around 1930, putting pressure on old-school eugenicists to calibrate a range of environmental factors into their 'human pedigree material' when assessing the likely causes of feeble-mindedness.

Hogben stated that three-quarters of mental defectives were the 'high-grade' feeble-minded, and 'it is from this section that all the striking and sensational cases are drawn'. He admitted that there may have been some inherited factor, but that it was their physical and social surroundings that had drawn this out. 'Sterilisation will indeed cure these social ills, as decapitation will cure a toothache.' Eugenicists' zeal, he wrote, meant that their argument had been lost before it began, because their claims 'have been vastly weakened in this country by gross over-statement motivated by class bias'. Hogben would go on to describe eugenics as 'ancestor-worship, anti-semitism, colour prejudice, anti-feminism, snobbery, and obstruction to educational progress.'

However, this criticism of traditional eugenical thinking, from within the discipline, only served to invigorate eugenicists. Historian Pauline Mazumdar describes the process thus: 'At this point in the early 1930s, with the tide of left-wing criticism setting against it, but feeling the fair wind of the sterilisation campaign in its sails, the eugenists felt themselves to be the focus of all that was vigorous in human biology. Their traditions and their problems were still at the centre of modern thought, even for those who were most determined in attacking the movement.'

The Brock Report itself noted that strong hereditarian evidence remained noticeable by its absence: 'We find a remarkable consensus of opinion among those who have had long experience of institutional work and of defectives in general that the proportion of defectives with certifiably defective parents is small . . . It is impossible in the present state of our knowledge about the causation of mental deficiency to forecast with certainty whether a child of any given union will exhibit mental abnormalities.' And then later, a little more ambivalence:

> We find ourselves compelled to the conclusion that in a large proportion of cases of mental disorder the prime aetiological factor is some inherited peculiarity, and that this peculiarity shows a strong tendency to be transmitted. If such transmission could be prevented, it is reasonable to assume that some diminution in the incidence of mental disorder would result.

This is a typical old-school eugenics non-argument: we have little evidence, but we 'feel' it to be true anyway and it will probably therefore be a good idea to proceed as though it were true.

That said, the official mind baulked at anything it considered eugenical extremism. Lord Eustace Percy, First Baron Percy, Conservative politician and Stanley Baldwin's President of the Board of Education, wrote in May 1929 that he had been horrified by the eugenic rhetoric and practice in many US states:

> I am afraid that I tend to be rather prejudiced on this subject by my experiences in America before the war when all 'progressive' Americans went mad on the problem of md and I have never quite recovered from my reaction against their exaggerations ... I have always been violently opposed to compulsory sterilisation, but I frankly don't quite know what my feelings are about the legalisation of voluntary sterilisation ... But you must let me think over the problem further before expressing an opinion.

Percy was writing to Lord Riddell, lawyer turned newspaper magnate and managing director of the *News of the World*. Born in relatively humble lower-middle-class circumstances, George Riddell's brilliance had seen him succeed in law and then as a press baron. He was the type of self-made, steeply upwardly mobile man who was over-represented in eugenics circles. Riddell had forwarded to Percy the lecture he had written in the spring of 1929 for the Medico-Legal Society, of which he was president, entitled 'The Sterilization of the Unfit'. In the twenty-page text of the lecture, and in a follow-up letter, Riddell repeated all the traditional unproven eugenic claims – like breeding like, a demographic emergency is upon us, etc., etc. Riddell quoted Harvard professor Edward East: 'We are getting a larger and larger quantity of human dregs at the bottom of our national vats.'

Riddell was furious that only doctors and policymakers were ever asked to contribute to eugenical debates, whereas accountants and businessmen would have pointed out the huge cost savings to be achieved by voluntary sterilisation, permitting more defectives to live in the community – much cheaper than institutionalisation. 'The truth is that we are spending far too much on these unfortunate people and thus penalising normal children upon whom the nation must depend for its future existence', he told Percy. He claimed that at a teachers' conference, a teacher had told him, '"Many of these children who come to us are like animals

– they grunt almost like pigs, but we do wonderful things with them after some years." I intervened, "At what cost?"'

To Minister of Health Neville Chamberlain, Riddell wrote a nonsensical statement, that 'it looks as if we are going to be eaten out of house and home by lunatics and mental deficients'. He said this was because it was proving difficult to ascertain and institutionalise them since 'mental defectives are extremely persistent and clever in eluding observation'.

Few, though, went as far as Dr Richard Berry, the medical director at Stoke Park Mental Deficiency Colony near Bristol, who wrote to *The Times* in February 1930 on the potential for a 'National Lethal Chamber ... When we have all seen for ourselves, as I have, the disasters which may attend the under-developed human brain, I feel confident that many will agree with me that it would be kinder to the human race to put some of its more chronic mental derelicts out of their, and our, misery.' This proposal caused an outcry, and undeterred, Berry wrote a long letter to the *Eugenics Review* expanding his vision:

> I do not share your views as to the 'sanctity of human life' or 'the almost insuperable legal and practical difficulties' which a lethal chamber would involve. There was surely little or no sanctity of human life in the War, and there do not seem to be any legal difficulties involved in judicial murder by hanging. By a stroke of the pen the politician condemns hundreds of thousands of his fellow-men to death as fodder for cannon, and the law can always take away the 'sanctified' life of the murderer. Why, then, should we be so anxious to preserve the life of the almost brainless, senseless, speechless idiots and imbeciles when it seems almost pathetic to condemn them to live their lives as helpless automata? Why spend, as I am informed England actually does, £93 per annum per head on such human refuse, and only £12 per annum per head on the normal, healthy child?

This shocked the *Eugenics Review*, and, increasingly self-aware that eugenicists must not come across as 'cranks', its editorial retorted: 'Extreme proposals always do worse than defeat their own ends; and we do not wish eugenics to run the risk of being regarded as anything other than it is – a reasonable, organic development of the policy of public health.'

REYNOLDS'S ILLUSTRATED NEWS

STERILISATION!
WHO ARE the UNFIT?
By HAMILTON FYFE

"Society" Means the Rich and Wrong

Self-Made Men are Always Dangerous

By J. Jefferson Farjeon
TALL TALE CUT SHORT
Alas, Statistics

Is Sterilisation Necessary?
MIGHT BECOME WEAPON OF THE CRANKS
First Steps to Eugenics
By GEORGE GIBSON

NOT FIT TO HAVE BABIES.
Peril of Having Dictators to Control Our Lives.
By GERALD GOULD.

Mr. Gerald Gould.

CULT OF THE UNFIT.
BIG FAMILIES WITHOUT A NORMAL CHILD.
STATE'S BURDEN.
STARTLING DISCLOSURES IN OFFICIAL REPORT.
Special to "The Sunday News."

But for many, 'cranks' as a descriptor came pretty close. 'Not Fit to Have Babies: Perils of Having Dictators to Control Our Lives' was the headline the *Sunday Chronicle* gave to an anti-eugenics article by influential columnist and reviewer Gerald Gould. 'Give an official the right to forbid, and he begins to think he has the right to command', Gould wrote, foreseeing the totalitarian aura that would shortly surround the sterilisation debates.

> Who are the 'unfit'? And who is to decide the question of fitness or unfitness? . . . Are we going to entrust the bodies of our fellow-creatures to officials in this connection? And if in this connection, in how many others? Where is the process going to stop? . . . The fact is that very, very little is known, or can ever be known, about human heredity . . . If the effects of breeding among racehorses could be calculated with mathematical accuracy, what would become of the bookmakers? . . . Who is so arrogant as to claim that the moral purposes of mankind can be attained by selective breeding?

This problem of definition – and who was to do the defining – was picked up on by H. Robins, writing in *The Catholic Times*. Robins pointed out the ongoing difficulty of distinguishing 'high-grade feeble-minded people' from 'the average citizen'. Daydreaming of a utopian pre-Protestant English past in which the less socially and intellectually capable were easily integrated into the community, Robins argued that 'mental deficiency as a problem affecting the state is of purely modern origin. The charity of the Middle Ages embraced without difficulty such defectives as could not be cherished by their families, and if there were any of "higher grade", the elastic social organisation found for them an easy means of frugal livelihood . . . Mental deficiency is largely coeval with the industrial era . . . Paganism once more solves its problems with the knife.'

He also attacked eugenicists for cherry-picking their 'illustrative' case histories and statistics, and accused them of religious bias, too, warning that poor Roman Catholics would find themselves targeted by the race-improvers: 'Catholics must fight most strenuously against its [eugenics'] acceptance, or England will soon be one large Mental Deficiency Home, with most Catholics as inmates.'

The left-leaning newspapers (for there were once such things in Britain) attacked openly on terms of class. Failed Labour Party candidate but successful journalist Hamilton Fyfe wrote in *Reynolds's Illustrated News* that there was a case for sterilisation – of the 'worthless, incapable, spendthrift' members of the aristocracy and upper class. Turning his gaze on the business community, Fyfe maintained his facetious tone: '"Self-made men" are especially dangerous to the community ... Their success has been secured (I am thinking of monetary success) by trampling on rivals by the use, very often, of methods which are near criminal ... To allow such anti-social qualities to be transmitted is most unwise.'

George Gibson, secretary of the Mental Hospital Nurses' Association, was one of the most outspoken trades unionists on the topic of sterilisation. He believed that the sex segregation of the feeble-minded had never been properly implemented. Despite the fearsome wording of the Mental Deficiency Act, efficient and thorough ascertainment of the 'feeble-minded' did not take place in a consistent manner – and never had. For the most part, this was because of financial retrenchment in all the public services; but Gibson pointed out that there had been large pockets of resistance to assessing children and young people, in the teaching professions and among social services staff. Gibson stated that 'segregation has not yet been seriously tried', guessing that some 240,000 people in England and Wales were likely to be feeble-minded but without ever having been ascertained as such. Gibson provided the official statistics for 1 January 1933, which showed 63,893 people had been ascertained and certified under the Act – almost equal numbers of males and females.

Gibson agreed with those who worried that voluntary sterilisation would be the gateway to something more sinister: 'Sterilisation of the unfit is an initial step towards selective breeding ... There is a very real danger that sterilisation, if permitted, would become a weapon in the hands of cranks.' He successfully proposed a resolution against sterilisation at the Trades Union Congress conference of 1934.

One other body critiqued the drift of the mental deficiency doctors in these years. The National Council for Lunacy Reform (NCLR) was, as its name states, a pressure group that sought to promote patient rights and hospital conditions in the mental illness sector. Formed in 1920, it had its roots in a very early mental health advocacy body, the Alleged Lunatics' Friend Society, founded in the late 1830s and succeeded by the Lunacy Law Reform Association from 1873. At its AGM of 1931, the NCLR laid its cards on the table: blind faith in doctors remained a problem and 'must be broken down. The danger today is not ecclesiastical despotism but medical domination.' Two years later, the NCLR focused its fire on the mental deficiency system, with senior member J.W.J. Cremlyn stating 'that it is essential to curtail the power of the Board of Control. This country would not tolerate Hitlerism and there are signs that it is at last dawning on the British public that we have, in our midst, an insidious dictatorship, far worse, owing to the fact that its effect is largely unseen. The workings of the Board of Control do not see the light of day and are a menace to a free people.'

A Conservative Party supporter (and future parliamentary candidate), Cremlyn was disgusted at his party: 'In spite of an overwhelming majority in parliament, the Conservatives do little or nothing to safeguard the liberty of the subject.' He was amazed that the government should allow such a state of affairs to exist. 'If the Board of Control so decree, a person may be detained indefinitely as mentally defective in spite of a hundred certificates by mental specialists to the effect that that person is normal! . . . One might as well appeal to a tiger in the jungle to release its victim.' With a very odd coda (for a Tory), he announced that 'the Socialist Party are today the custodians of national liberty'.

***

Membership of the Eugenics Society in these years had grown to around 650–700, and it retained its reasonably deep pockets and long reach into the establishment. It had started its campaign of targeted propaganda to legalise sterilisation in the late 1920s; after Brock, it now launched another well-funded lobbying programme to persuade parliamentarians and the British public to back moves to introduce voluntary sterilisation for the feeble-minded. When a Ten-Minute Rule Bill was proposed on 21 July 1931 seeking leave to present such a measure, it was defeated by 167 to 89.

Before the General Election of 1935, the Eugenics Society-backed campaign contacted all 544 parliamentary candidates, of whom 202 expressed sympathy towards the measure. Conservative Minister of

Health Sir Hilton Young remained unpersuaded by the Society and its supporters – astutely aware that showing support could alienate swathes of the working-class and Roman Catholic voters. Perhaps it was having one eye on the electorate that led Labour leader at the London County Council Herbert Morrison to reverse the LCC's initial support for voluntary sterilisation measures. Morrison stated, 'We have no right to take liberties to experiment upon people not in a condition to make a decision for themselves.'

Historian Greta Jones has traced the Labour MPs who in the early 1930s were sympathetic to voluntary sterilisation, and has concluded that the handful or so were almost without exception on the right of the party, having formerly been Liberal, and one defecting eventually to the Conservatives. However, just like first time around, a significant number of politically active women on the left went against Labour Party and trades union policy and supported voluntary sterilisation. The same impulses that had motivated female support for eugenics before the First World War were still at play: that control of fertility was crucial to improving women's lives – and so sterilisation was viewed by some as an extension of contraception, which the organised left (with the exception of Catholic Labour voters) fully supported. In addition, many campaigning women retained the mistaken belief that vasectomy would act to curb excessive and aggressive male sexuality that led to rapes and sexual assaults. As various women's groups across the country passed resolutions backing voluntary sterilisation, it was common to see statements linking such a procedure to a lessening of male sexual violence, particularly with regard to offences against children. The Eugenics Society funded a phoney grassroots organisation in 1935, the National Workers' Committee for the Legalising of Voluntary Sterilisation, having recruited Dr Caroline Maule, an American-born physician who was a Labour Party member, to set it up, and featuring three other Labour members. The Committee only lasted three years, and Maule was warned never to attempt to pass off voluntary sterilisation as Party policy. Nevertheless, at the 1936 National Conference of Labour Women, she moved the following resolution: 'That this conference, believing it is wrong to condemn people to a choice between unnatural abstinence and the risk of bringing into the world unfit children who will be a misery and a burden to society, favours the legalising of voluntary sterilisation . . .?' She was seconded by Dr Edith Summerskill, who described the operation as 'preventive medicine', and reassured that there would be no question of compulsion. Miss C. Maguire opposed, saying that the Brock Committee had simply discovered what it had been required to discover, and calling out the 'ceaseless propaganda' of eugenicists. But the motion was carried by a large majority.

Every debate on sterilisation between the wars referenced eugenic activity in other countries, as though seeking context on how 'reasonable' British race improvers were. So that's what we'll do now, too.

15

UNNATURAL SELECTION: WHAT WENT ON ABROAD

'On the edge of the beautiful Valley of the Moon' (as she described it) stood the Sonoma State Hospital – California's main institution for the feeble-minded. Cora B.S. Hodson, general secretary of the British Eugenics Society, came to Sonoma in 1933 as part of her international sterilisation fact-finding tour. 'The utmost possible freedom obtains, and little groups of boys and girls move to and fro from one set of buildings to another, joking and playing as readily with the staff as with each other'. Bright sunshine, glorious surroundings, and the head nurse turned to answer Hodson's query: sterilisation of the youngsters 'had been so long the general practice that it had become a tradition'. It was freely discussed within the institution, the nurse told her, and the inmates actually saw it as a reward – a preliminary to obtaining their release into the community. Hodson soaked this up. It was everything she had wanted to hear. In the infirmary wing, bedbound newly neutered defectives told her of how proud they were of having had the procedure. On the question of consent (which she said had been worrying her), she was told by senior staff that it had taken quite a lot of work to persuade the 'ignorant' and 'low-grade' Mexican, Spanish and French families of the defectives to agree to the operation, but it was worth it, because it gave 'ideas of social value to people with only a glimmering of what enlightened citizenship and parenthood means'. And California was already seeing a reduction in expenditure on welfare payments for the unwanted children of the poor.

California had been the third US state to legalise the sterilisation of its undesirables (in 1909). By the time of Hodson's visit, the total number of people in American institutions who had been sterilised stood at 16,066 – 9,067 females and 6,999 males. Of these, fifty-three per cent had been inmates in Californian facilities – the Golden State was way out in the lead. One-fifth of America's sterilised were individuals who had been certified as mentally defective, and four-fifths were mentally ill. There had, by 1933, been no wholly independent review of the results of the sterilisation programme – either for California or for the rest of the US. For what it's worth, a questionnaire sent out by British mental welfare workers in 1951 received the following brief responses from Californian doctors: 'Patients who subsequently married have shown considerable resentment – occasionally a precipitating factor in producing neurosis.' 'Some ex-patients are reported to be unhappy and disappointed to a marked degree.' 'Relatives sometimes report that sterilisation exposes a girl to undesirable attentions from the opposite sex.'*

The Brock Committee noted that the Californian sterilisations had clearly not been performed in order that patients could return to the community – Cora Hodson had been misinformed. Discharge and retention rates showed that many remained in detention: of the sterilised mentally ill, forty-seven per cent of men and twenty-nine per cent of women were still held in mental hospitals after the operation, while with defectives, thirty-four per cent of men and twenty-eight per cent of women remained institutionalised. 'In our view, there is no justification for sterilising defectives who are unfit for community life', Brock reported.

Hodson eulogised California as the birthplace of all that was modern and scientific – its fine Observatory and two more under construction, cosmic rays, IQ testing, and the Human Betterment Foundation. Launched in 1928 in Pasadena, and taking inspiration from the New York-based national movement the Eugenics Record Office, the Human Betterment Foundation had its major financial backing from citrus fruit businessman Ezra Seymour Gosney. Like the British Eugenics Education Society, the American Eugenics Record Office and the Human Betterment Foundation lobbied hard for the introduction of sterilisation of the insane and the feeble-minded.

---

* See Appendix 3 for US states' responses to the 1951 questionnaire.

But California had been twenty years ahead of the Human Betterment Foundation: in 1909 its first sterilisation law legalised compulsory sterilisation of inmates of its state mental hospitals and homes for the feeble-minded, as well as of selected convicts in prison. An amending act of 1913 made parental (or a guardian's) consent a condition of operating upon a feeble-minded patient. A second amending act detailed tighter measures required to prove the hereditary nature of the condition suffered by the person to be sterilised.

The mentally ill, rather than the learning-disabled, bore the brunt of California's hard-line approach. Cora Hodson drove, on a gloomy day, to the Stockton State Hospital – California's largest mental hospital. Here, assistant director Dr Margaret Smythe ('a most charming and sympathetic woman') performed two sterilisations in front of Hodson. It was for their own good: 'They are made to realise that, however poor and feeble, they have a part to play for the good of mankind by voluntarily allying themselves with the endeavour to prevent inherited misery in the future', Hodson parroted.

As we have seen in chapter 6, Indiana was the first American state in which sterilisation was legalised, in 1907; by 1933, some 217 sterilisations had been completed in Indiana. The state of Washington was the second to legalise and had sterilised thirty habitual criminals by 1933. In addition, some thirty US states had passed laws banning certain types of individuals (including the feeble-minded) from marrying, on eugenic grounds.

Federal government overturned many US states' sterilisation legislation, on the grounds that it went against the Fourteenth Amendment, pledging equal treatment of all citizens; and, with regard to the neutering of criminals, on the grounds that this constituted 'cruel and unusual punishment'. However, the landmark 1927 lawsuit Buck v. Bell found that the state of Virginia's sterilisation of eighteen-year-old Carrie Buck, deemed to be feeble-minded, had been justified. Buck had been raped, disbelieved, and was committed to the Virginia Colony for Epileptics and Feebleminded in Lynchburg as 'a moral imbecile', since she was pregnant and unmarried and her IQ test gave her a mental age of nine. Carrie's mother and sister, Doris, were also in the Lynchburg colony as feeble-minded. Doctors examined Carrie's daughter Vivian when she was seven months old and decided that she had the 'look' of low intelligence, they said.

Carrie Buck with her mother, Emma, photographed by the Eugenics Record Office.

Sterilisation for Carrie was arranged, but her court-appointed guardian challenged this. As part of the legal case, the Eugenics Record Office examined the family pedigree and declared that the Buck women 'belong to the shiftless, ignorant and worthless class of anti-social whites of the South.'

The Supreme Court upheld the decision to sterilise Carrie by eight to one, and she was operated upon on 19 October 1927 as Virginia's first legal overtly eugenic (rather than therapeutic) sterilisation. Carrie went on to marry twice and died age seventy-six, judged to have 'normal' intelligence and behaviour by those she lived amongst. Carrie's sister, Doris, was sterilised in 1928, and discharged two years later. She married and only found out many years later why she and her husband had been unable to have children; the Lynchburg superintendent eventually told Doris what had happened to her. In an interview she gave in 1980, Doris said: 'I never knew anything about it. I'm not mad, just broken hearted is all. I just wanted babies bad . . . I don't know why they done it to me. I tried to live a good life.' (Vivian Buck did well at school and showed no signs of any 'backwardness', but died aged eight in 1932 of an intestinal complication arising from a measles infection.)

The Buck v. Bell judgment came from Justice Oliver Wendell Holmes, who wrote the opinion: 'It is better for all the world, if instead of waiting to execute degenerate offspring for crime, or to let them starve for their imbecility, society can prevent those who are manifestly unfit from continuing their kind. The principle that sustains compulsory vaccination is broad enough to cover cutting the Fallopian tubes. Three generations of imbeciles are enough.'

After this Supreme Court ruling, sterilisation was given the green light across the country and by 1935, twenty-eight states were sterilising their undesirables within institutions (twenty-four of them also permitted compulsory procedures). Three thousand people from the various undesirable categories were being sterilised each year, slightly over half for being mentally defective.

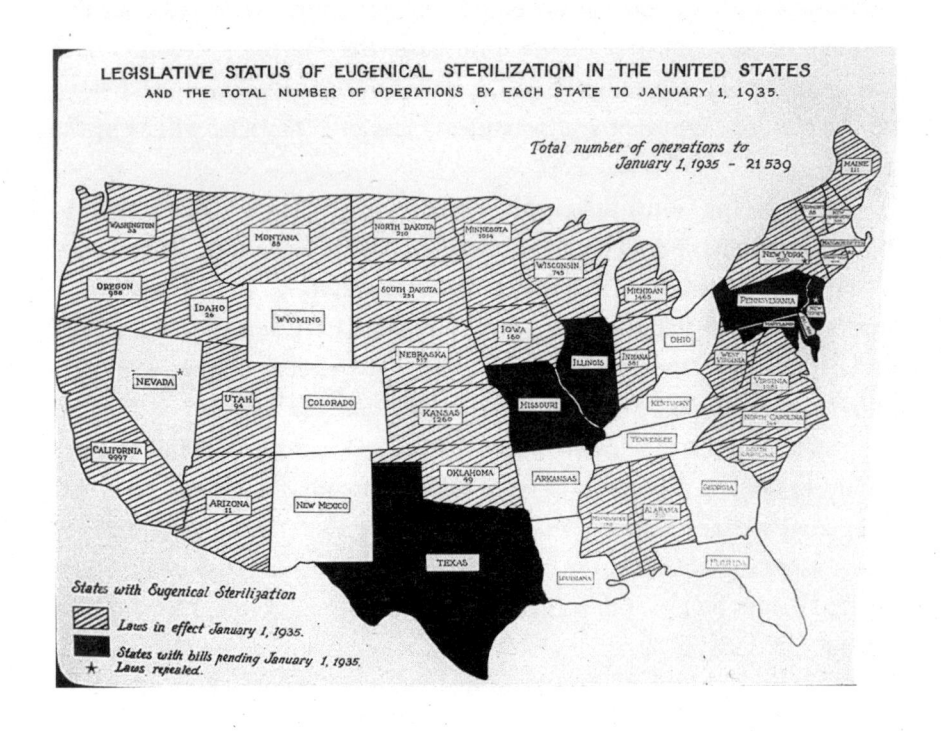

Women were much more likely than men to be sterilised, and almost all were poor. For what it's worth, Superintendent Bell of the Lynchburg Colony (as in Buck v. Bell) opined that sterilisation was a measure that protected women, who were easy prey to the sexual aggressions of males of superior intellect, as well as those of her own mental level ... The feeble-minded male cannot enter into serious competition with the normal male for the affections of the feeble-minded female ... The female defective is, generally speaking, more dangerous eugenically than the male ... It is therefore evident that if all mentally defective women were sterilised, there would be but little reproduction of feeble-minded persons.

As we have seen, non-WASP minority groups were significantly more likely to be sterilised, and in Virginia, almost half of the sterilised were black. Although it falls outside the scope of this book, the horrifying escalation in many US states of the sterilisation of Afro American people on purely racist grounds continued into the 1970s – within and without the medical and penal settings in which sterilisation originally took root.*

Eleven US states additionally permitted castration (disproportionately carried out on less-well-off males, and on black men). So did Denmark – the second-most-cited foreign country after America in the late 1920s and early 1930s, when the British were ruminating on sterilisation of the feeble-minded. Of a population of 3.5 million, Denmark had 28,000 ascertained feeble-minded people – or one in four hundred Danes. It was generally believed that of all nations, Denmark had undertaken the most thorough ascertainment, and the country was looked to as a model of no-nonsense, pragmatic responses to its undesirable population. The Danish government passed its sterilisation legislation in 1929 (the earliest in Europe), initially targeting sex offenders and the psychotic within psychiatric institutions; the new law permitted both compulsory sterilisation and compulsory castration.

In 1923, 100,000 women had petitioned, via the Danish Women's National Council, expressing concern at an alleged rise in sex offences – they wanted the authorities to impose much stricter penalties for such crimes, and ideally castration for repeat offenders. Certain medical experts

---

* Contemporary works on these more recent sterilisations in America include Alexandra Minna Stern, *Eugenic Nation: Faults and Frontiers of Better Breeding in Modern America* (2015); and Edwin Black, *War Against the Weak: Eugenics and America's Campaign to Create a Master Race* (2003, new edition 2012).

agreed with them, but ultimately the public prosecutor would not back the measure. Nevertheless, by the time the sterilisation law came into effect, in 1929, castration was seen as acceptable. That said, by 1945, seventy-eight per cent of the sterilised were in the feeble-minded category, with females twice as likely as males to have been operated upon.

In terms of segregation, a tiny island had been commandeered by the internationally famous Kellersk Institute. Sprogø is halfway across the Great Belt strait and from 1923 it housed – for indefinite stays – hundreds of girls and women who were declared feeble-minded on grounds of being promiscuous, or who were repeat runaways, from 'social problem families', or who were simply deemed unable to look after themselves. They worked for no pay on the farm on Sprogø, and the facility's founder, Dr Christian Keller, himself made the decision on if and when a woman could leave; none could do so without first being sterilised. The last woman left the island in 1961. For males, Livø island, in the middle of the stretch of water called the Limfjord, was chosen by Keller in 1911 to incarcerate the morally deficient/sexually aberrant, alcoholics, vagrants and epileptics. It, too, closed in 1961.

The Danish sterilisation law faced little sustained popular opposition, though teachers worried that in IQ testing their pupils they may have been making them vulnerable to certification as mentally deficient and thus to being sterilised. In the 1940s, it became clear that a number of the sterilised had turned out to be 'late developers', with no form of mental disability at all, and a formal inquiry was convened. Compulsory sterilisation in Denmark ended in 1967.

Sweden passed its first sterilisation Act in 1934 and between 1935 and 1948, some 15,519 operations took place, of which 12,108 were upon women. The majority of these were women deemed to be mentally defective but the second-largest group were 'women exhausted from many childbirths'. In name, this was 'voluntary' sterilisation; in fact, doctors would press ahead with the operation if consent was proving problematic because the patient was either too mentally ill or too learning-disabled.

In the 1930s the remit was extended to include those Swedes with 'social maladjustment' or with an 'asocial disposition', such as vagrants, prostitutes, or those who were deemed to be avoiding the world of work; all could now be sterilised. In 1941 the definition of mental deficiency was expanded again, to incorporate those who could be described as having 'an anti-social way of life'. Compulsory sterilisation in Sweden ended in 1975.

***

And so at last we come to the spectre that hovers over the entire subject of eugenics and which altered its course and its reputation for ever. The slow realisation of the rhetoric and then the actions of the Third Reich with regard to disposal of its undesirables modified many of the debates and disputes in Britain from 1933; and for those still asleep at the wheel, the liberation of the death camps dragged eugenics into a very harsh spotlight. After 1945, anyone advocating the artificial improvement of human stocks would need to defend themselves in light of the extremes to which Germany's Nazi regime took the notion of the perfectibility of humanity.

Before 1933, German eugenics embraced a broad spectrum of politics: liberals, socialists and conservatives all worked in the field, and many Jewish scientists were to be found within the discipline. Germany's post-First World War struggle to get back on its feet economically placed a new focus on the cost of maintaining its 'unproductive' citizens – including its mental defectives. One report put their number at between eight and ten per cent of all Germans between the ages of sixteen and forty-five. The expense of their maintenance troubled many during the Weimar Republic, and in 1932 Germany's first draft sterilisation law was written – comparatively late in the day, when set alongside several US states and Denmark. The draft proposed a voluntary procedure, with proof being required that the defective traits were actually hereditary. It was welcomed in most quarters, with the Catholic Church being the only firm and consistent opponent. However, it never became law, because in January 1933 Adolf Hitler came to power – with a very different approach to slashing welfare budgets and ensuring that only the healthy bore children. He added a new 'racial purification' element to Germany's eugenics debate; and under Hitler's twelve-year dictatorship, eugenicists who wanted to carry on working in the discipline now had to align themselves (either overtly or tacitly) with the extremism that would lead to the extermination of millions of people. Jewish scientists were now forced out, under the 1934 statutes requiring professionals to prove they had Aryan ancestry.

The Nazi government, in July 1933, passed its law on mandatory (not voluntary) sterilisation of those who – in the opinion of the newly created system of Hereditary Health Courts – were suffering from: feeble-mindedness; schizophrenia; manic depression; hereditary epilepsy; hereditary blindness or deafness; the neuro-degenerative condition Huntington's disease; chronic alcoholism. Much of the inspiration for the Nazi sterilisation programme came from observing how the US states of Virginia and California had implemented their laws. The admiration was sometimes mutual, with Leon Whitney, executive secretary of the American Eugenics Society, stating in 1934 that 'many far-sighted men and women in both England and America have been working earnestly towards something very like what Hitler has now made compulsory . . . They have fought courageously and steadily for the legislation of what they consider a constructive agency in the betterment of race.'

Under this Law for the Prevention of Hereditarily Diseased Progeny between 200,000 and 400,000 Germans were compulsorily sterilised after examination by the Hereditary Health Courts between 1934 and the outbreak of the Second World War. More than half of the operations are thought to have been performed upon the feeble-minded, while the mentally ill were another large contingent among the neutered. Within the community, authority figures took increasing part in sniffing out the undesirables among them.

Much more iniquitous – off the scale in terms of human barbarism – than compulsory sterilisation was the Nazi euthanasia campaign, with an estimated 7,000–10,000 children and young people being killed, usually by lethal injection, sometimes by starvation. The earliest murders were conflated with the concept of 'mercy killings' – to spare the child suffering and to relieve parents of their financial and psychological burden. Also, patriotism was appealed to – the Fatherland needed to allocate its resources carefully as war loomed: do the right thing, and volunteer yourself to be put down if you are unhealthy – and your unhealthy offspring too. Many of the earliest killings did in fact take place with the acceptance by the parents of chronically disabled children that they were unlikely to survive whatever 'treatments' were offered. Welfare agencies often simply compelled single parents to hand their children over.

Hitler extended the programme to include adults as war was declared, on 3 September 1939 – the maniacal hatred of perceived weakness was now coupled with an urgent need to free up beds in institutions, in order to accommodate the likely large number of military casualties. The policy was named *Aktion T4* – after the address of its headquarters, the villa at number 4 Tiergartenstrasse in Berlin. From here was orchestrated the selection and extermination (by gassing) of over 70,000 long-stay patients – the mentally deficient, the mentally ill and the physically disabled or chronically sick. The government's financial savings resulting from these murders were plotted on to graphs.

Who did this? Historian Michael Burleigh has described the personnel of *Aktion T4* as an 'odd assortment of highly educated, morally dulled humanity' who set about recruiting 'a staff of people willing and able to commit mass murder.' Former students were recommended, and the SS provided

> seasoned hard-men who could cope with the physicality of mass murder. Teams of these people were despatched to six killing centres. The doctors who monopolised killing were given perfunctory briefing sessions in Berlin, and then gradually inducted into murder, progressing from observing the procedures to carrying them out themselves. Most of them were quite young, socially insecure and hugely impressed by major academic names... that is, the usual accompaniments of petit bourgeois academic ambition.

The gassing programme paused in August 1941, and the Holocaust got under way – the destruction on racial grounds of millions of Jews and over half a million Roma. The murder of the mentally and physically disabled carried on, but at a slower rate, and now by starvation and injection rather than gassing, and branching out to include the elderly in care homes, plus vagrants, foreign forced-labour and people injured in Allied bombing raids. These murders carried on right through to the Allied victory and the fall of the Third Reich.

The vast majority of members of the Eugenics Society and its associated commissions were revolted by the racism of Nazi eugenics, when they sent investigators over in 1934 to forge links and make comparisons with the discipline in Britain. They could not have known about *Aktion T4*, because with the outbreak of the Second World War, relations between British and German scientists ended.

But we *are* able to consider British views of German sterilisation plans for the feeble-minded between 1933 and 1939. 'Tarred with the same brush' was a powerful fear among British eugenicists when Hitler's proclamations were backed up with legislation, from July 1933. The *Eugenics Review* of that month broadly welcomed Hitler's plans for compulsory sterilisation, but inveighed against his racialisation of the subject – describing as 'doubly deplorable' his wish to eliminate 'foreign' blood in Germany, and passionately attacking Hitler's anti-Semitism (his 'race warfare') in particular. However, the same editorial took the opportunity to say that the Führer should 'reduce', 'restrict' and 'subordinate' the power of the Roman Catholic Church, with its 'anti-scientific obscurantism'.

The *Eugenics Review* was clear to put distance between the compulsory nature of Nazi sterilisation and the Eugenics Society's wish for a voluntary measure.

> From pulpit and platform, in the press and also doubtless in parliament, they will warn the trustful British public that in supporting the Society's policy it will, in fact, be unwittingly taking the first step down the slippery declivity that leads to compulsion, bureaucracy and the tyranny of racial or social majorities.

Cora Hodson was one of two breakaway members of the Eugenics Society. She was, from 1932, the chair of the International Federation of Eugenic Organizations (IFEO), which hoped that Hitler's implementation of eugenic ideas would act as a spur for compulsory sterilisation in other countries. The IFEO's membership, as its name declares, was drawn from all over the world, and around fifty members attended regular conferences. In terms of British engagement with Hitlerian eugenics, Hodson was joined by enthusiastic racists George Pitt-Rivers, anthropologist, and Reginald Ruggles Gates (geneticist and formerly Mr Marie Stopes). Nevertheless, the Eugenics Society considered the IFEO a somewhat embarrassing sideshow and, in Britain at least, it never had clout.

British eugenics thus pulled back in time to be able to salvage some remnants of political acceptability and respectability when, in 1945, the full truth of Hitlerian eugenics became known.

In 1910, Liberal MP Charles Masterman had told Winston Churchill in a memo on the topic of sterilisation, 'I feel inclined to suggest minuting this "Bring up again on Jan 1st 1950". I am absolutely certain that any proposition whatever giving these powers is outside of the possibilities of the present legislation . . . We cannot carry public opinion with us beyond this.' When 1950 came, public opinion had shifted significantly further against such measures.

It is a banal but nevertheless true observation that the enormity of the Third Reich's crimes is difficult to take in. This book turns away now from those horrors (inadequate word) to continue with the far less terrible events in Britain, with regard to the allegedly feeble-minded.

16

THE EAST CHALDON VICARAGE AFFAIR: THE 'FEEBLE-MINDED' IN THE COMMUNITY

Peering through his binoculars, a resident of the Dorsetshire village of East Chaldon spotted an adolescent girl secreting herself within a large hedge in the grounds of the vicarage, 'hiding like a hare from the hounds', he later said.

This wasn't the oddest thing locals had seen with respect to the girls who lived there. Long-time resident Farmer Cobb said that he had seen 'maidens' fleeing the vicarage on four separate occasions, and on others had heard loud crying from within the house. 'Something out of the ordinary is going on there', said Cobb. At a tea party given by author Llewelyn Powys for new arrivals in East Chaldon, novelist Sylvia Townsend Warner and her lover, poet Valentine Ackland, discussion turned to 'the oppressed servant girl at the vicarage' who had twice that day tried to escape. The set-up at the vicarage was notorious in the village, the tea-party guests told Sylvia and Valentine.

The pair did not let their blow-in status deter them, and they marched over to the vicarage and demanded entry. They were allowed in by Mrs Katherine Stevenson, sixty-six, and her daughter, Joan, thirty-seven, who then allegedly told their angry visitors that 'practically all the girls had sexual mania' and so could not go anywhere unattended, and that they also 'had immense physical strength'. Sylvia shouted, and Valentine shook her walking-stick at the Stevensons, prompting Sylvia to think, regarding her new lover: 'Righteous indignation is a beautiful thing . . . I watched it flame in her with severe geometrical flames.'

Not long after this clash, the Stevensons' Great Dane bit Sylvia's chow. This may well have been the final straw, and in mid-May 1934, Sylvia and Valentine wrote and circulated a petition within the village, garnering forty-four signatures. It read:

> It is our considered opinion that neither Miss Stevenson nor Mrs Stevenson are suitable persons to have the care of mentally deficient girls, who, we would suggest, should be treated with sympathy and understanding, and not subjected to too rigorous discipline. In view of the numerous complaints which have been made as to the manner in which this home is and has been conducted, we would strongly urge that the whole case should be thoroughly investigated by the Dorset County Council. We feel sure you will see that pending such investigation it would be advisable that no steps should be taken that would involve a renewal of their lease or retention of the vicarage in any other form by those persons as tenants.

Sylvia Townsend Warner, left, and Valentine Ackland.

When the council told Whitehall of this petition, the Board of Control sent an inspector, Miss E.G. Colles, to the vicarage to investigate. But the Stevensons would not let her in and said the identity document Miss Colles brandished might be fake. Miss Colles reported of the older woman, 'I found her thoroughly truculent ... She refused to accept my official card ... She looks a most unpleasant person and I should be reluctant to leave feeble-minded girls in her hands.'

Six girls between the ages of fifteen and eighteen were being boarded out at the East Chaldon vicarage. When tested, all were declared 'subnormal' by the specialist (male) doctor next sent to the village by the Board of Control. (This time the Stevensons were fully compliant.) He spoke to each of the six separately, and all but one claimed to be happy. He was pleased to hear this, as he felt their situation was not a desirable one for a young woman to be in. He reported that they were living 'in this very isolated spot with little or no amusements and not enough to occupy their time.' Trips into Dorchester (eight miles away) were arranged for them only twice a year; and despite the 'muddled', 'untidy' and even 'dirty' nature of the large house, they clearly were not being kept busy. They spent their daytimes in a recently constructed 'very comfortless shed extension' at the back of the building, measuring 15ft by 14ft, and they slept in a dormitory scarcely bigger. The Stevensons received £1 a week for each girl they took in, plus a clothing allowance, on the understanding that the girls would be taught housewifery skills so that they might try their luck in the outside world and perhaps eventually even find a job in domestic service, thereby becoming self-supporting. However, it was clear to the doctor that no such training was taking place. And no graduate of the vicarage had ever actually found a position.

The Stevensons were approved to take in these cases by the Dorset Voluntary Association for Mental Defectives ... of which Mrs Stevenson herself was the secretary. She claimed that magistrates and doctors had visited the vicarage no fewer than twenty-eight times in recent years, and that they all approved the set-up. In legal terms, the Board of Control inspector could only really fault the Stevensons on a technicality: they had not used the correct forms to notify the Board of the admission of the most seriously learning-disabled girl, Ruth P—, from rural Oxfordshire, aged eighteen but with a mental age of eight. She was the only girl who 'looked unwell, depressed and perhaps afraid', the doctor reported. 'Mrs Stevenson informed me several times before I saw this girl that she was so nervous I would get no answers out of her. I found no difficulty getting answers when the questions were down to her mental level ... In my experience, anyone with experience with defectives should recognise the mental defect here.'

The Board judged the Stevensons to be entirely inappropriate for the role of caring for and training the girls, and all were removed from the vicarage. Mother and daughter launched a libel case against Sylvia, Valentine, Llewelyn Powys and Farmer Cobb – claiming that their petition had been motivated by malice. Through their lawyer, the Stevensons also maligned their former residents as 'sometimes the daughters of tramps, sometimes of criminals, sometimes of mental deficients. They included girls in unfortunate circumstances, and girls with terrible habitual tendencies and terribly vicious habits.' The Stevensons won their libel case and were awarded £175 in damages – a result that brought Sylvia and Valentine close to bankruptcy.

***

Section 51 of the Mental Deficiency Act permitted this type of boarding-out – and, in fact, ever since the legislation had been passed, much greater use had been made of 'guardianship' and 'supervision' than had been intended. Despite the impassioned rhetoric of the early twentieth century on the need to segregate defectives into colonies in order to prevent breeding, the sheer cost and the time required to create such institutions meant that care within a community setting presented itself as a significantly less expensive option. (As mentioned previously, the Act became operative just sixteen weeks before the start of the First World War.) By the 1920s, some began to wonder: if it had been so eugenically crucial to prevent mental defectives breeding, why had the resources not been made available for the institutionalisation of all those who were ascertained?

The concept of heredity had also come under pressure with the new understanding of how certain diseases could impact on the developing brain – including meningitis and epilepsy; and how head injuries could lead to changes in behaviour and personality. More importantly, a modification in thinking about mental deficiency came with the discovery, in the mid-1920s, of encephalitis lethargica – colloquially named 'sleepy sickness'. The syndrome remains mysterious but has widely been believed to be one of the after-effects of the Spanish flu/Great Influenza outbreak. The global pandemic that killed millions of people worldwide, from 1918, also impacted severely upon the health of many who survived it. Clinicians and those working in mental deficiency made a correlation between encephalitis lethargica and a deterioration in the intelligence level and behaviour of children who had not been born 'defective'. One physician wrote, in 1926, of children who had previously been well behaved becoming, after contracting encephalitis lethargica, 'bad-tempered, violent, vicious, irritable, quarrelsome, destructive, cruel, untruthful, dull, stupid, dirty in their habits . . . Thieving is common among them, but it is impulsive, rather than intentional.' An estimated seventy per cent of children endured psychological disturbance after recovering from the disease, with over one-third exhibiting behavioural and even criminal behaviour.

In light of this development, the government prepared a second Mental Deficiency Bill, in 1927. Josiah Wedgwood loudly opposed it ('Compulsory segregation put the erratic at the mercy of the fanatic, among whom none was worse than the expert philanthropist'); but it passed easily. One of the problems with enforcing the 1913 measure was now overcome – the reluctance of doctors and medical officers to diagnose deficiency if they were unable to trace a patient's heredity.

The 1927 legislation replaced the 1913 wording, 'a condition existing from birth to early age', with 'a condition of arrested or incomplete development of mind existing before the age of eighteen years of age, whether arising from inherent causes or induced by disease or injury'. (The 1927 Act also replaced the term 'moral imbecile' with 'moral defective'.) Given that, now, some defectives did not need to be prevented from breeding, since their condition was attributed to an environmental factor, the authorities could feel a little easier about the number who were living out on licence.

***

As mentioned in chapter 9, a number of 'voluntary associations', operating with the agreement and ultimate oversight of Whitehall, were in some parts of the country crucial to the placement of 'defectives' and monitoring their circumstances. As the years passed, an ever-larger number of ascertained defectives were billeted in the towns and villages of Britain. It's an aspect of this nation's provincial life that has largely passed unexamined. In each community, there were youngsters who had to abide by a curfew and who were explicitly prevented from forming attachments with the opposite sex; but also were policed for signs of homosexual relationships. These parallel lives were lived alongside 'normal' society – a shadow world of 'guardianship' and 'supervision', with the person out on licence often seen as 'odd', 'simple', sometimes a figure of pathos, sometimes someone to be made fun of, to be exploited, to be bullied. The English village in which I grew up in the 1970s was two miles from a colony, and I would occasionally overhear adults muttering darkly about such things as someone being 'not quite right', someone who'd been 'put away', someone 'a bit funny'.

In 1929, over 44,000 people who had been declared mentally defective were living under guardianship, statutory supervision or voluntary supervision; a further 26,000 were in colonies or other forms of institutional detention. By 1947 these respective figures were 72,000 and over 54,000. Increasing use was being made of hostels, some attached to colonies, others run by local authorities or voluntary associations, and operating as 'halfway houses'. Ascertained defectives who were living out in the community included those who were felt to be not so problematic as to require institutionalisation; others were out on licence from an institution to see if they were, after all, capable of living with some degree of independence; others still were simply waiting for a place in an institution to become available.

The local authority executive officer was the figure with the ultimate say-so on their fate. He liaised with his own Mental Deficiency Committee and with guardianship societies, completed the necessary documentation and would often personally accompany the youngster to an institution if an infraction of the guardianship rules had been spotted.

In East Chaldon, locals had become concerned for the well-being of the girls billeted with the Stevensons. However, the well-being of the communities themselves became problematic, too, with regard to the guardianship phenomenon. In 1932, the Brighton Guardianship Society was criticised by the local council because of the number of cases the

Society boarded out on the brand-new North Moulsecoombe council housing estate. The housing sub-committee expressed its concern that if so many undesirables were being placed on the estate (the majority of them coming from far outside the locality), this could damage the area's image as a place of relaxation, leisure and fun. According to the council's Mental Deficiency Committee: 'The importation of steadily increasing numbers of defectives into Brighton was felt to be a grave menace to the amenities of the town as a health resort, and the council were determined to take steps to limit the practice, even if this necessitated the promotion of a bill in parliament.' But taking in a Brighton Guardianship Society case provided a much-needed additional income stream in these years of high unemployment, and, given the general shortage of suitable foster parents, it seemed like a win-win – less well-off people were paid money to provide a home to youngsters who were generally proving hard to place. (In East Chaldon, the Stevensons had never hidden the fact that the £1-a-girl fee paid to them was a major source of their income.)

The social workers and Board of Control inspectors who monitored those living under guardianship or supervision came in for all sorts of criticism – the tabloid newspapers found a great source of stories in the figure of the snooper, intruding on someone's private life and family life. Popular Sunday newspaper *John Bull* took up the cudgels on behalf of young Bertie Brewster – his plight made all the more toothsome to the reading public because of his enrolment as a police constable in the Ipswich force. He had spent four years in the Royal Eastern Counties Institution at Colchester, certified as mentally defective because of his aggressive behaviour towards other children, and a subsequent seven years out on licence, first with his own family and then, because of tensions with his stepfather and his brothers, lodging nearby with a female guardian. Seven years of having his movements tracked, as part of his licence conditions, had built up fury within him. 'The present restrictions are handicapping my social life in every way', he wrote to the Board of Control on 23 February 1931. These were the rules of his life – he must:

Never be out of doors after 8pm.

Never change his abode without official sanction.

Never leave Ipswich – not even for a weekend or a holiday.

Never go to a kinema [sic] unless accompanied by his parents.

Never enter licensed premises.

Never have any women friends outside his family.

Never marry.

Bertie's former employer (for three years), J.S. Copsey, brushmaker, wrote in support of the young man who, he stated, was a good worker, conscientious, intelligent and didn't drink or smoke. 'I consider it a great injustice that you still keep him tied by petty restrictions', Copsey told the Board of Control. 'I cannot stand by and see him constantly nervous and worried through the thoughts of continued unnecessary interference with his private life.' He was being 'hounded down whenever he tries to help himself up in the world.' P.S. 'He has nothing to do with the opposite sex.'

John Bull, in full Gothic mode, recycled tropes and purple prose from Victorian madhouse dramas ('After lingering for years in the ghastly asylum . . .'), and the newspaper was additionally enraged that it was 'women mental welfare workers' who 'followed him in the streets' and called regularly upon his relatives and employer. Imagine that: women telling a man what to do.

The publicity worked. The Board of Control instantly rescinded Bertie's mental deficiency certification, and in an internal memo admitted that this should probably have happened years earlier.

The *News of the World* came in to bat for Marjorie B—, who while out on licence formed a perfectly innocent friendship with a young soldier during attendance at her local church services. It took just one week for somebody to notice this and report it to the council's Mental Deficiency Committee and Marjorie was whisked back into Etloe House in Leyton, East London – a Roman Catholic institution licensed to be used as a mental deficiency colony for females. She would spend a further seven years here.

There was nothing wrong with Marjorie's intelligence (or behaviour) in any case. Her story is typical of many who found themselves under lock and key for years. Her parents had an unhappy marriage and the household was stormy. The rows frightened little Marjorie, and when she was still very young, teachers noticed a decline in the quality of her schoolwork. As a divorce loomed, Marjorie was placed into a special school, aged eleven. Here, she fell even further behind, because the lessons were only in handicrafts and domestic work. When she was tested she was on the one hand found to be extremely shy and nervous, and on the other, contemptuous at questions she considered fit only for a four-year-old – she refused to answer. She was marked down as ineducable and sent to a residential school for the 'backward'; and then at age sixteen she passed straight into Etloe House. Here, she worked in the laundry and the dressmaking workshop. At the age of twenty-two she was allowed to live out on licence and 'it was while I was there that I met a young soldier. He was kind to me and took me to the cinema twice. I remember he was tall and had dark hair. He told me he had no mother.'

When the *News of the World* ran her story, the Board of Control permitted new tests, plus a psychiatrist's report. These showed Marjorie, now thirty-three years of age, to be intelligent and capable, and, despite twenty-two years in captivity of one sort or another, she exhibited no emotional problems. The tasks she was unable to complete satisfactorily during this testing session were all the result of her having had no academic education after the age of eleven.

The Board freed Marjorie from her certificates immediately. She said she had entirely lost her once strong religious faith, having experienced the over-reaction to her friendship with the soldier and the speed with which someone, or some people, within the congregation or the ministry had misinterpreted the friendliness between herself and the young man.

Peggy B— also met a soldier while out on licence from a mental deficiency colony. Knowing that her licence conditions forbade even a friendship with a member of the opposite sex, Peggy and the soldier kept their love affair secret. She became pregnant and while her lover was away fighting in the Second World War, she gave birth – having run away to her mother's home. When the baby's father was demobilised he came to find her and asked her to marry him. Three days before the wedding was due to take place, news reached the authorities of the impending marriage. Somebody must have informed on them. An ambulance arrived and Peggy was dragged screaming into it, back to the colony to which she had originally been committed.

Once again, the tabloid press came to the rescue: thwarted romance, a baby left motherless, cold-hearted authority, local gossips – all the stuff of fantastic front pages. This won Peggy a conditional release, although not to her family home. She was re-tested and found to have no educational or (amazingly) psychological problems. However, the press reported that while Peggy now had her freedom, she had 'lost touch with her baby's father. Thanks to the authorities, it remains an illegitimate child.'

Peggy's back story was that she had been certified under the Mental Deficiency Act as a moral defective after committing two petty thefts. Her father, briefly home on leave from the War, signed the documents for her detention in a colony, thinking that this was to be a short spell in a kind of training home. Once demobbed, he discovered that he was unable to get her out.

In Mildred M—'s case, the young man she had met and fallen in love with asked her to marry him. They both wanted to be open and honest about their relationship, and so he wrote to the medical superintendent of the colony from which Mildred was on licence. He told the superintendent that he could offer Mildred a stable home and that both his income and his prospects were good. The van came for Mildred within days.

At Eaton Grange hostel in Norwich for women living out on licence, requirements for residents to be, and to appear, demure took the form of a list of Rules of Behaviour. The matron insisted: 'You are not allowed to speak to MEN or BOYS or have anything AT ALL to do with them. You are NOT ALLOWED to use LIPSTICK or POWDER, but you must always be CLEAN and TIDY.' Dr Sheena Rolph, who has written the history of Eaton Grange, found that 'the women's lives were monitored, and their freedom in a very major way was curtailed. No relationships with men were permitted.'

The voluntary Ladies Committee at Eaton Grange was involved in surveillance, as were the matron and the local Mental Deficiency Committee. 'They interviewed women, warning them about their behaviour.' That said, Rolph also saw much foresight in the matron's largely successful attempts to help the women forge relationships with the outside world, mainly through work placements and organised leisure pursuits. Rolph uncovered 'the depth of some of the friendships that grew between people in the neighbourhood and the women from the hostel', even though they were forbidden to develop friendships with men.

In the Blofield Hall hostel for males, seven miles outside Norwich, friendships also developed with workmates at the local tin factory, and when factory workers offered the hostel men lodgings with their own families, many Blofield men were at last able to achieve full discharge. Norwich City Football Club would come to the hostel to play against the men there, in the early 1950s. However, attempts to integrate hostel men into the local social clubs proved more challenging, as the clientele often 'did not know how to relate to the [Blofield] men ... so it was difficult'.

***

Many certified people who were out on licence were vulnerable to workplace bullying – because of their supposedly being 'mental' or 'retarded'. Mrs S— revealed her worries about her sister E— who was about to be allowed out on licence:

She knows nothing about work and the outside world. She is so shy that I feel she will be put on by work mates, as I myself have seen many dirty tricks played on one who has been a little dull . . . and I feel the same thing will happen to E— unless I can keep her at home and take her about until she gets used to civil life. We all love my sister very much and neither of us want E— to go back there [the colony] again ever, as she has never known what love or outside life is for thirty-one years, and she is forty-one now. So please help me so that we may try to make up to her for some of the love she has never had.

E— had been 'put away' at the age of eleven, back in 1928; now, three decades later, she was to have a glimpse of the outside world, thanks to the relentless campaigning of her sister. Her father had signed the necessary paperwork in the belief that E— was to be sent to a 'special school' for additional catch-up education – for six months.

People on licence could also find themselves implicated in socially unacceptable, anti-social or criminal events. A study of the Somerset mental deficiency authorities revealed that the main reasons for patients' recall to an institution included 'homosexual practices', 'associating with the opposite sex', 'seen talking to a schoolgirl', 'an incestuous relationship with her father', 'associating with undesirable men' and 'interfering with little girls'.

It was against the law to have sex with a mental defective, and the burden of proof (from 1913) fell on the accused to prove in court that he hadn't known that the female with whom he had had intercourse had been certified under the Mental Deficiency Act. Voluntary association staff and social workers were scrupulous in checking the cleanliness, cubic space and physical comfort of the homes to which inmates were to be billeted; and the 'respectability' of the foster figure was crucial, too. But assessing for potential sexual predation proved problematic. The judge at the Sussex Assizes trial of Henry Cottingham, in July 1927, spoke out about the inadvisability of giving the care of a female certified under the Mental Deficiency Act to a woman who had an adult son living at home, while she herself was out all day at work. The Brighton Guardianship Society had placed a nineteen-year-old woman and (they later told the court) claimed that they had insisted that she should never be allowed out and about on her own. However, the Society had not realised that she would be spending the entire daytime unchaperoned and unprotected in the cottage.

Cottingham found himself in the dock for having intercourse with this certified defective. He claimed that he had had no idea that the nineteen-year-old was any such thing. He said he had never heard her described in such a way, nor seen any paperwork in his mother's cottage to that effect. He also underlined his War service and his good character. This worked, and the judge discharged him.

Violet A— was licensed to the care of her own father in December 1936. Robert Taylor, thirty-six, was Mr A—'s lodger at the house in Hastings, Sussex, and when the following summer Violet gave birth to a child, her father accused Taylor of having taken advantage of the girl. Taylor strongly denied this. But when she gave birth to a second child, eighteen months later, her father contacted the police. Taylor claimed that Violet had consented to sex with him.

A second man, William Hammond, fifty-one, was a frequent visitor to the A— household. He said he had asked Violet to marry him and she told him that she would, once she was free of her certification. Hammond denied being the father of her second child and took the opportunity to claim that a number of male visitors regularly attended the house – any one of them could have been responsible.

Hammond and Taylor were both arrested and three months later pleaded guilty at the Sussex Assizes, in July 1939; the former was placed on probation for two years, and the latter was sentenced to twelve months in prison.

George Bostock was sentenced to twelve months in prison when the jury did not accept that he had been ignorant of the mental deficiency of a twenty-five-year-old he had been having intercourse with for several months. Bostock, thirty-seven, a one-eyed newspaper vendor from Lancaster, told the court he hadn't known 'she was mental', just that she was 'a bit backward'. He had fully intended to marry her, he said. Despite his disability, he had served in the Pioneer Corps during the War, he told the court. The judge liked this, and said that he personally would be inclined to deal leniently with Bostock. However, he said, letting him off lightly might send a signal to other men that this was not a serious matter, and so he jailed him.

John Lawson had been sent to Rampton as a moral defective following a series of violent criminal offences committed before he was eighteen. His behaviour seemed to stabilise at Rampton, and so after eight years, he was transferred to Calderstones colony in Lancashire. In July 1936 a panel of doctors decided to assess Lawson's ability to live independently by allowing the twenty-nine-year-old out on licence for a month; Lawson's brother had requested this, and promised that he would be able to provide close supervision of John. However, the dossier detailing the extent and savagery of his original crimes had not been in front of the panel when they made their decision; these included sexual assaults upon boys and young men, robbery with violence, housebreaking and larceny. Two weeks after his temporary release, he knocked on the door of Bertha Holgate, a stranger to him, and asked if he could have a cup of tea. Pitying his appearance (she had thought he was a tramp), Mrs Holgate invited him in, made him tea, offered him cake, and also set about mending his torn clothing. While her back was turned, he picked up a piece of wood and battered her around the head, fracturing her skull.

At a subsequent court hearing, Mrs Holgate was awarded £3,500 personal injury compensation in her civil case against Calderstones and the panel who had allowed Lawson out on licence. The prosecuting counsel, Sir Patrick Hastings, inveighed against the local authorities of Britain, who needed to be replaced entirely, he said. They were responsible for 'this dangerous creature [having been] put into the world without reasonable care'. Lawson was a 'moral delinquent, or defective, with strong criminal or immoral tendencies which would never disappear'. Although Lawson's brother had pledged that he would undertake careful supervision of John, in fact, the entire Lawson family was out at work during the daytimes, leaving John totally unmonitored. Sir Patrick continued that 'the paperwork that had granted him licence was wastepaper ... The lunatic was under the supervision of nobody.'

***

After the Second World War, the issue of the labour exploitation of 'defectives' was the subject of even more intense discussion and campaigning than in the 1920s and 1930s. When Labour won the 1945 election, the Attlee government repealed anti-trades union laws, and union membership rose to 9.5 million. Union leaders and members made their voices heard to a much greater extent than before in discussions on social and economic matters and how best to reconstruct the postwar country. The heavy workloads for pin money within mental deficiency colonies – undertaken by the 'high-grade' defectives – has been discussed in chapter 8. These same concerns around cheap labour were expressed with regard to patients out on licence. In 1950, the Society of Labour Lawyers and the Socialist Medical Association wrote to the Ministry of Health pointing out that 'the possibilities of exploitation are obvious, because the employer always has the weapon of threatening to send the employee back to the institution to hold over his or her head . . . Employers can exploit because they have the power to return the mental defective to the institution.'

Only rarely were union-agreed rates paid to defectives out on licence, and in farming, there was no agreed Agricultural Wages Board minimum pay, nor limitation of working hours. The worst excesses happened when the patient was licenced directly to the care of an employer, rather than to a family or a foster carer. Living-in arrangements by employer-guardians could be scandalously inadequate.

Gangs of the certified were taken on to perform labour-intensive physical work such as fruit-picking and other types of harvesting. This occurred ostensibly while they were out on licence but there was a suspicion that the licence periods coincided nicely with Mother Nature's cycles and the need to reap her bounty in certain summer months. Farming was one sector where overwork and underpay were rife with regard to defectives out on licence. In the late 1950s, the National Union of Agricultural Workers (NUAW) and the Transport and General Workers' Union (T&GWU) alerted the Ministry of Agriculture, Fisheries and Food of around forty young males out on licence in farms across Kent and Sussex. Arrangements had been made for the boys and men directly between the colonies in which they had been detained and the farmers who needed labour; the local voluntary association rubber-stamped these agreements, but appeared to undertake little due diligence on pay-rates, hours of work, conditions and medical care. Sometimes only bed and board were supplied at the farmsteads in return for hard labour in the fields. The NUAW had found boys and men on licence working seven days a week, without any overtime payment. The union additionally brought up the topic of how much more dangerous postwar farming had become: was it wise to have mentally deficient people operating the new range of machinery that had become so common across the late 1940s and 1950s? Between 1951 and 1955 there had been 698 fatalities among agricultural and forestry workers – should patients be working in such an environment? An NUAW meeting debated this topic in the winter of 1956 and decided they should not.

The two unions' concern triggered a number of surprise visits to the Kent and Sussex farms by Board of Control inspectors. The Brighton Guardianship Society, which had placed the patients with farmers, made the rather weak defence that, regardless of a weekly wage, working in the countryside was 'likely to restore many of these folk to full mental health – this was, to the Society, of more importance than any money they might earn . . . The Society's main aim was the rehabilitation of as many of these unfortunates as possible, and their return to normal life'.

The T&GWU wasn't at all impressed by this argument: 'It seemed that the type of farmer who was asked to have these defectives on his farm was not altogether altruistic in his intentions.'

***

Wartime played a strange role in the life of Alma S—. Aged twenty-five, and an inmate of St Catherine's Certified Institution near Doncaster for nine years, her licence was revoked after nine months, in 1934, because the lady in whose home she worked as a maid had caught her 'gossiping'. The lady did not like gossip. A second licence period proved much more successful, with Alma working as one of four maids at Loversall Hall, a mansion not far from St Catherine's. The superintendent abruptly terminated Alma's licence period – but only because he now wanted her to work as a nurse orderly within the facility. The wealthy owners of Loversall Hall contested this, writing to the Board of Control in 1941 to point out that licences for inmates were not granted and revoked according to the manpower needs of the institution. 'She is to be brought out of freedom and into captivity. This being the case, it may well have a permanent effect on her state of mind, and all the good work of years thrown away', they wrote. The superintendent of St Catherine's openly admitted to Alma's employers that but for the fact that the country was at war, he would have Alma discharged entirely from certification – there was nothing much wrong with her. At Loversall, Alma had shown herself to be quiet, docile, a useful worker, with occasional episodes of obstinacy the only black mark against her.

The Board of Control was not sympathetic to the owners of Loversall Hall. Why on earth did they need four domestic servants anyway – didn't they know there was a war on? The Board quite liked the idea of Alma and others like her filling the staffing gap at St Catherine's, with the proviso that inmate-nursing orderlies should never have sole charge of any patient.

In the event, the episode showed how ridiculous it was to detain Alma at all – clearly a competent person and a good source of labour in the national emergency. She let it be known that she didn't actually want either the nursing job or to continue as a maid at Loversall Hall; and the Board of Control instantly discharged her. Within a week, she had three job offers to consider.

A group of lawyers decided that the licence system 'operates as a sieve which sifts the wrong way. The higher the mental standard of the patient, the more likely he or she is to find the ludicrous living conditions impossible to be kept.' That group of lawyers worked for the National Council for Civil Liberties, and it was the NCCL who at long last forced the overturning of the mental deficiency laws.

PART 5

LIBERATION?

17

'A SCANDAL ON A FAR
FROM SMALL SCALE'

The National Council for Civil Liberties had never, in its thirteen-year history, found itself so frequently stonewalled and fobbed off as in its dealings with the Board of Control. Founded in 1934 to publicise and campaign against the aggressive policing of demonstrations and marches, the NCCL subsequently took on a greater number of powerful interests in the United Kingdom – highlighting such matters as press censorship, racial discrimination and police harassment of striking miners. In 1947 it turned its attention to the failings of mental health legislation, and in particular the injustices of the Mental Deficiency Acts. During this fight, the Board of Control simply refused to address crucial arguments, with not even a 'courteous appreciation of points raised and a reasoned statement of reply', according to the NCCL.

The nation would be shocked, the organisation continued, when it realised that most people who had fallen within the remit of the mental deficiency machinery had no history of criminal or anti-social behaviour; and a huge proportion had IQs that placed them within the 'normal' range.

It all started with a series of angry letters from 'Windy Ridge', a house just outside Newton Abbot, Devon. In 1947 retired accountant George Scott-Rimington believed he had stumbled upon a network of corruption and graft, with the regional hospitals board, a mental deficiency colony and members of the local magistracy acting in concert to certify and retain youngsters who had either fallen behind at school (and so might be termed 'educationally backward') or who had exhibited perfectly normal adolescent misbehaviour, such as running away, disobeying parental demands or petty thieving. Scott-Rimington's most serious accusation was that money was skimmed off by the managers of the colony, and paid out to cronies. Officially, the colony was a private 'not-for-profit' company; in reality, certain individuals invoiced 'for services rendered' and received payouts. 'It is an old stunt', he wrote to the NCCL, 'and any skilled company accountant will explain it to you.'

Scott-Rimington and two Devon-based solicitors who had tried to investigate what looked like a huge provincial stink – with national ramifications – claimed that they had, as a result, been threatened with attacks upon their businesses as well as upon their personal integrity. Scott-Rimington's health began to decline rapidly because of the anxiety this had caused, he claimed.

The young woman whose experiences had triggered his investigation would be known under the pseudonym 'Jane'. Her family had begged Scott-Rimington not to have her (and their) real identity made known, and so that is what I shall call her too.

'Jane' was certified in July 1943 as a mental defective on the order of a local magistrate, subsequently discovered to be one of the company directors of the colony. Jane had been in a special school. As Josiah Wedgwood et al. had warned thirty-five years earlier, Jane, like hundreds, and perhaps thousands, of special school pupils, was rubber-stamped across to a mental deficiency colony when she reached sixteen. 'Educational backwardness' was not supposed to be conflated with mental deficiency – that had been made clear in the 1913 Act; all too often, that is exactly what happened.

It is impossible to know what tests Jane was set during her assessment, or how she responded: repeated requests by Jane's family and local supporters to see the medical evidence upon which she had been declared defective were refused. Worse, someone had forged the parental consent required for Jane's detention as a certified mental defective – her father, a widower, would later take an oath to swear that this was not his signature on the form. He could easily prove that he had been stationed with the RAF in Bedfordshire in July 1943 and was then sent overseas from there to fight in the War.

Jane's family and supporters fought for years for an official re-examination – a process that the creators of the 1913 Act had deliberately ensured was as difficult and expensive as could be, according to campaigners. Scott-Rimington alleged that officials had intimidated all Jane's character witnesses before her hearing and had forbidden them to offer her support during the procedure, in 1948. At this hearing, Jane was so terrified she could barely answer the questions put to her by the magistrates. This event was 'a farce', wrote Scott-Rimington, as the JPs permitted no evidence to be heard against the original certification. Jane was returned to the colony. She was a hard worker. She once had all her hair shorn as a punishment, she said; occasionally they would inject her with something and beatings were common. Now that she had witnessed these cruelties, there was no way they would permit her to leave – or so her supporters suspected.

Jane had one legal challenge left – extremely rarely invoked but permitted by the legislation: a direct appeal to the Lord Chancellor. In Scott-Rimington's words, 'We met with the same strange attitude of obstruction.'

The new Labour government had inspired in the retired accountant only scorn. Minister of Health Aneurin Bevan had treated his letters 'with derision', giving only one 'fatuous' reply. The Ministry pursued a 'slow-motion performance of writing bland answers to serious complaints'. He hoped that Bevan et al. would soon be thrown out of office. Scott-Rimington claimed that while he believed in the brand-new National Health Service, he felt that it was not being democratically run, and was instead controlled by 'secret committees, who always have one eye on the voting list'. The moment it became apparent that Jane's case was going to prove problematic for them, officialdom began to show 'complete ruthlessness', even 'megalomania'. They closed ranks, and then they began to lash out. Where in the statutes had that been sanctioned?, he wondered.

The solicitors managed to win Jane a period out on licence to her father and stepmother. But the local authority's Mrs Orchard allegedly came to the house often 'and adopted a very abusive and threatening attitude when I refused to answer questions about my private affairs', stated Jane's stepmother. 'I have had seven long years of being worried and harassed by officials.'

During her time on licence, Jane and her supporters collected fifty-four statements from locals to the effect that she had never shown any indications of having either low intelligence or abnormal behaviour.

What Scott-Rimington wanted to happen in future was for the parental consent order to be signed not before a magistrate but before a Commissioner in Oaths – who would ensure that the parent knew exactly what mental deficiency certification entailed for their child. All the expert medical testimony should be open for the family, and other supporters, to consult. They had to know precisely what evidence had been used to label their relative, or friend.

Too often, a parent was either too ignorant or busy or felt too intimidated by an authority figure to ask about the nature and length of the detention. The procedure as it stood was faulty, since there were simply too many examples of parents believing they were handing over their child for a finite spell in a residential educational establishment with extra rules and regulations to curb wild behaviour. 'Humble working-class people' should not have to spend money and rely on the goodwill of 'busy professional people' to help them out in a battle against public authorities, Scott-Rimington said. (But here, he had more or less described the very dynamic between the NCCL and those who turned to it for help with mental deficiency cases, as we shall shortly see.)

In July 1949, his health declining and his worries about persecution on the rise, Scott-Rimington handed Jane's case over to the NCCL, and one of their most successful campaigns was born. His final highfalutin flourish, as he laid the scandal before them: 'My associates and I are very happy to feel that our own profound belief in the liberty of the subject is shared by other people . . . in the eternal fight of mankind against intolerance.'

As a result of the NCCL's agitation, Jane was released from certification, and blossomed. She found new friendships and pastimes and caught up on her learning now that she was *outside* the system; in total, she had experienced a decade of no proper academic education. According to Scott-Rimington she exchanged the 'wooden, monosyllabic attitude that I have seen in the men who have spent five years in a prison camp' for a 'charming' manner. She got a job as a cashier within days of gaining her freedom, and she married two years later.

It had been Scott-Rimington's strong impression that the British public were apathetic on the subject of mental deficiency miscertification and the exploitation that many of the certified experienced. But the fact that newspaper editors knew that such stories sold papers suggests he was mistaken in that belief – as does the flood of letters that the NCCL received following its announcement that it was to throw its weight behind the subject. Over two hundred letters arrived from parents, relations or simply concerned parties regarding miscertification and retention of inmates.

Sifting through the case histories taken on by the NCCL from the late 1940s to the early 1960s, several themes emerge. Firstly, it is striking that many girls or young women appear to have been dumped into the mental deficiency system for want of any other place for them. This was no longer even tacitly being done to prevent the breeding of the unfit: the eugenical dimension of the mental deficiency laws seems to have faded away. Nevertheless, the urge to incarcerate 'wild girls', and thereby prevent the birth of illegitimate children and so decrease the number of so-called 'problem families', is almost certainly behind this phenomenon. As with the failed attempt to introduce sterilisation, it didn't have to be a strictly eugenical measure – more a means of preventing the 'wrong' type of person becoming an inadequate parent, whom the state would have to bail out in some way.

Trivial incidents triggered certification for a number of the girls and women; and in the NCCL correspondence it sometimes seems that adolescent girls were taking the blame for more general family dysfunction. Glenda E— was bullied by her eldest brother when he was home from the Army, including a physical assault. But it was Glenda who was deemed to be problematic and she was re-called to an institution. She had originally been sent to a colony in Yorkshire at the age of fifteen when the epileptic fits that she had suffered since the age of eight meant that she could no longer do her job, working in a hairdresser's salon. The Hospital Management Committee wrote to Glenda's mother in June 1952 to say they were sorry that the girl had had to return 'but I think it was best, for somehow you appear to be a family that does not get on well together and clearly this is not good for a young girl. Licence at home for your daughter has, therefore, proved to be a failure.'

Her mother replied that Glenda 'has done no wrong, and as I visited her on Saturday . . . she was very unhappy and said she had been taken back without warning . . . I don't know what my elder boy wanted to provoke her at all for. Seeing he is an organist and a local preacher, you would have thought he would have loved his only sister better than that.'

In her battle to have her daughter returned to her, Glenda's mother canvassed the neighbours and put together their comments for the NCCL to use in its battle with the hospital's management. 'I have never known her to be a bad girl – it is terrible to say she isn't normal', said one neighbour. 'I have always known her to be a quiet normal kind of girl . . . never inclined to run after the opposite sex', said another.

Glenda's mother added that things 'at home are more peaceable [now that] her two exasperating brothers are away'.

Her mother visited every month, and Glenda was distressed that she couldn't come home. Her mother claimed she 'was told they were keeping her there because she is a good knitter . . . She is my only girl and ought to be a comfort to me now her brothers are away in the Army. She is a lovely girl. It's heartbreaking when I have to leave her.'

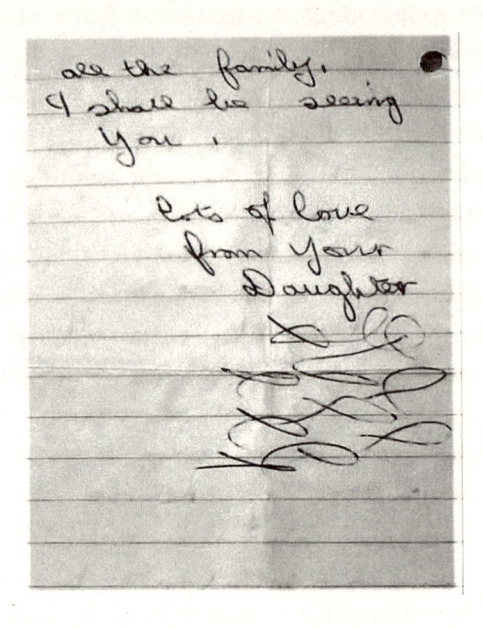

all the family.
I shall be seeing you.

Lots of love from your Daughter
xx

Glenda would never get the chance to experience adult life outside the institution. In 1953 she either jumped, fell or was chased over the balcony banister of an internal stairwell, falling several feet and breaking her back. She died five-and-a-half months later of her injuries. The coroner was not happy at the conflicting explanations of what happened, but the case was not pursued.

***

We have already seen how an unhappy home life and divorce could cause a child to underperform at school; 'special school' was created as a solution to this, but in fact it often tended to lead to mental stagnation and the consolidation of educational 'backwardness'. One petty theft, one incident of insubordination, one unsuitable boyfriend, one flash of temper – each could now precipitate detention for years when added to a documented background of educational problems. The mother of Jean P— turned to the NCCL in desperation in 1953. Jean, nineteen, had spent twelve months in St Lawrence's Mental Hospital at Caterham, in Surrey. Her parents had divorced when she was eight years old, and she had then been evacuated during the war – with this added disruption, she fell way behind in her schoolwork and was sent to a special school. Here, staff described her as 'lazy' and 'obstinate' but not lacking intelligence. At the age of fourteen, she was left alone in her mother's house with an acquaintance of the family. The next day her mother noticed she was 'extremely nervous and frightened' and when her mother and aunt asked her why, Jean said she had been sexually assaulted by the friend of the family. The police refused to press charges as there was no corroborating evidence.

When nearly eighteen, on the way home from a dance, Jean was subject to an attempted sexual assault. Again, the police would not act as there was no corroboration. After this, her mother found Jean difficult to control. After a row, Jean stormed out of her mother's home, in May 1952, and stayed with a succession of friends and neighbours. Jean sent a friend round to collect her belongings but her mother would not hand them over. So Jean waited until her mother was out and entered the house, removing not just her own clothing but also a coat belonging to a lodger – the young women had frequently swapped and shared garments, and so this was not an intentional theft on Jean's part. Nevertheless, she was sent before the magistrate on a charge of stealing. Truculent, she refused to answer when asked if she knew the difference between an orange and a potato, and which is the longest river in England. Jean was certified mentally defective and bundled off to Ward 223 of the Caterham institution.

The Board of Control directed her supporters to the fact that Jean's paperwork showed that she had been tested in 1944 and found to have a mental age of eleven; however, in 1944 Jean *was* eleven, as the NCCL would later triumphantly point out. They added that 'the evidence would suggest that the girl has suffered a good deal of emotional upset due to the break-up of her parents' marriage, the six years spent in a residential school, intensified probably by the attempted assault at the age of fourteen. Such a condition is a condition of maladjustment, and not of defect. It is not subject to the Mental Deficiency Act.'

Mrs P—'s letters to the NCCL reveal a second common feature – a crushing feeling of powerlessness on the part of the families. Doctors, colony superintendents, local authority officials, the Board of Control seemed to assemble themselves into a high wall when requests were made for re-assessment, or a period of licence, or an investigation into allegations of brutality or neglect within an institution. Mrs P—'s solicitor put it this way: 'This was a perfect example of petty dictation in the lives of working-class people that a bureaucrat sometimes develops'.

That the NCCL responded thoughtfully and – crucially – kindly, was appreciated by those who turned to them for help.

'They are so big and hard to fight . . . I know you are trying to do a lot for these girls, and I know that you have to be wary too . . .'

'Hoping I shall soon be able to really thank you for all you have done for us both. Please excuse my mistakes in writing as my little boy has just got up out of bed and is very impatient . . .'

'Do please get her out before her seventeenth birthday, I shall never let her go again. God meant us to have life freely, not shut up like they are. She was broken hearted when they came for her.'

'No wonder there are so many broken hearts today in trying to endure. I shall never forget her being taken back in that ambulance. She was just like an angel. O what could I do? Write to me all the best advice you can.'

'I have spent pounds to get her released and have only been laughed at by all of them . . . I felt like tearing the whole place apart, wishing I was a Samson. Oh well, I shall not bore you about the Concentration Camp any more but in any case, win or lose, I shall not forget you have tried to help . . . Respect to all of you who are trying to do the right thing.'

The parents and relatives stressed that the person in need of rescue could easily prove that they were a useful member of society – hard grafters, never looking for a handout, clean, decent: 'socially efficient', to use the technocratic jargon of the day. The majority also made reference to their belief that the real reason for the ongoing detention was how very hard the inmate worked in the institution – scrubbing, laundering, knitting, mending, helping feed and move less able patients.

With an estimated quarter of people detained in colonies having an IQ of over seventy, and prison costing twice as much per inmate as a colony, the NCCL believed that much of the injustice they were uncovering was indeed financial in origin (as examined in chapter 8). The Council's secretary, Elizabeth Allen, reported that seventy-five per cent of inmates were 'quite capable of adapting themselves to a normal life, and this is clearly demonstrated . . . by the satisfactory later histories of many who have been released after a period of detention.' Allen wrote that the government simply needed to accept higher running costs of institutions, and release their hard-working captives.

Parsimony may also have been behind the decision to place a youngster in a mental deficiency colony since non-residential 'schools for the educationally subnormal', created by the 1944 Education Act, had spaces for only 15,000 pupils, where the true figure in need of such schooling was estimated to be 28,000. *The Lancet* pointed out, in 1951, that a good, progressive idea was failing because the government was not making enough money available: 'Insufficient educational facilities are being provided for this type of child, and some of them are inevitably reported as "ineducable". If their home conditions are bad, they may then [instead] be certified as mental defectives and ordered to a mental deficiency institution.' As we have already seen, this was never supposed to happen: as far back as 1913, performing badly at schoolwork was hugely different to being 'defective'. But for thousands of children, it had led to years of institutionalisation.

The NCCL believed that certain magistrates had indeed been abusing the permission granted within the Mental Deficiency Act to declare that a parent was 'unreasonably withholding consent'. They also uncovered examples of clearly forged signatures from a mother or a father who had died before the committal proceedings. 'Many and devious are the means employed to obtain consent', the NCCL decided.

In less than forty years, a bureaucracy had established itself in this field, interpreting the law as it pleased and answerable to itself alone, stated the NCCL of the mental deficiency machinery set up in 1913 and barely modified since then. In no case where they had obtained a patient's freedom did the Board of Control or the Ministry of Health ever offer an apology and certainly never an offer of compensation.

In 1951 the NCCL published its findings to date in a forty-page book entitled *50,000 Outside the Law*, an immensely readable digest of anonymised case histories and satisfyingly catalogued areas in which the law was wrong and needed to be changed. It was a pithy, lucid *j'accuse*, and caused an instant stir – high sales meant that it had to be reprinted within a year. The NCCL's mental deficiency case load now rose to six hundred.

The NCCL doubted that there were many unimprovable cases to be found in the colonies. Five out of six cases in which the NCCL called in independent experts were found not to be mentally defective after all.

What had been instituted was 'preventive detention', the NCCL argued. The recently passed Criminal Justice Act permitted the courts to order preventive detention of a maximum of fourteen years; this was for a known recidivist ('persistent') offender who was undergoing a sentence after conviction for a crime. With the Mental Deficiency Act, however, youngsters were being held for decades. The NCCL cited 'Bert', who had been detained as a 'moral defective' following a conviction for burglary and was now entering his twentieth year of incarceration in a colony. But 'John' beat even Bert's epic confinement. Placed into an institution in the north of England at the age of three, he was discovered by the NCCL forty-four years later – illiterate, having never been given any schooling while in captivity, and unable to state any particulars about his past except for his date of birth. When a supporter asked to see the paperwork by which John had been certified (as a toddler), the institution simply took John before a local magistrate and claimed that he could be described as 'found neglected'; one of the clauses in the Mental Deficiency Act that could trigger a certification was for someone to be declared 'abandoned, neglected or without visible means of support' and therefore in need of being sent to 'a place of safety', i.e. a mental deficiency institution. The magistrate asked no questions and John was now returned to the institution – the documentation, though totally false, retrospectively making his detention appear legal. The NCCL dubbed him 'The Forgotten Man' in their magazine, *Civil Liberty*, and stated that 'we found a situation we would have thought impossible in the twentieth century'. There is a happy ending, though. The NCCL found John a job in a hotel and reported that he 'gives every satisfaction ... He has opened his own Post Office savings account, he has begun to develop personal friendships, he is learning the elements of writing, the use of the telephone and 101 other little things that have been kept from him for forty-four years.'

A similar case, in which the 'found neglected' clause was misused, was that of Kathleen Rutty.* Kathleen had been in the care of Essex County

---

* Kathleen's real name is used here, as her case created a great deal of newspaper publicity and questions in parliament.

Council from the age of three months, and at age seventeen was transferred across to St Michael's Hospital in Braintree. Here she had a job, and was paid a weekly wage. In June 1948, aged twenty-three, she passed from St Michael's into the Royal Eastern Counties Institution in Colchester as a mental defective, under the 'found neglected' clause. This was clearly untrue – Kathleen had secure board and lodgings within the hospital, plus she was on their payroll. She also had £15 in a Post Office savings account. Six months later, she was allowed out on licence to the care of her stepbrother, in Clapton, East London, and quickly found a factory job. But the Royal Eastern Counties Institution wanted her back.

The NCCL assisted Kathleen in bringing a habeas corpus case, which the broadsheet newspapers followed keenly. Having John Platts-Mills QC and Lord Chief Justice Denman appearing in the same case would always attract readers who enjoyed the thrust and parry of the courtroom. Denman declared that he had looked over the questions that had been put to Kathleen to test her mental acuity and said, 'I could not have answered some of them myself.'

But what really swung it was the realisation that the creation in 1948 of the National Assistance Board by the Attlee government meant that no UK citizen could legally be defined as 'without visible means of support'. The local authorities now had a statutory duty to create a safety net – which had clearly been done in Kathleen Rutty's case; therefore her certification had been based on a falsehood. She was not 'found neglected' and therefore had not been in need of 'a place of safety'. Not only did she have a paying job (taxed and with National Insurance deducted) in the hospital, she also had a home, as a resident worker.

The Lord Chief Justice took the opportunity to state that he doubted the whole category 'borderline high-grade mental defective'; only psychiatrists would believe in such a phenomenon, he said, and even they disagreed with each other about it. 'No persons of any age are to be confined in institutions merely because doctors and officials thought it would be good for them.' The magistracy, he continued, appeared to have got it into their heads that they were to act as a 'rubber stamp' for doctors' and officials' decisions.

Frank Haskell, one of the chief organisers of the NCCL's mental deficiency campaign, said the Rutty case highlighted the 'wickedness' of the transfer of people from 'care' to mental deficiency institutions 'under the pretence that they had been found neglected.' He said that some 7,500 people were currently detained under the 'found neglected' pretext, and in every case, no local authority had produced paperwork proving they were neglected. He called for the release of everyone detained under that clause and for 'proper rehabilitation machinery' so that long-stay colony inmates could be returned to 'normal life'. Haskell reported that he had met one man detained as neglected at the age of eighteen who had only won his release at the age of fifty-three. Upon testing, he was found to have an IQ of 130.

Regarding the Rutty case, John Platts-Mills wrote that

> the false statements made by the authorities were routine and conventional in such cases and that the NCCL uncovered a great wrong. After the threat that many more legal actions would follow, the eventual outcome was that over five thousand such women were released. The Attorney-General, Reggie Manningham-Buller, told me that as many as ten thousand young women came under the Rutty decision.

Some 800 people detained under the 1913 Act were immediately released in 1957, and 1,000 the following year. John Platts-Mills, at the end of his long life, decided: 'Of all those I have fought, I think that the Rutty case gives me the most satisfaction in having secured a just freedom for the greatest number of people in the shortest time.'

18

THE KEYS TO THE DOOR

On 22 October 1953, Prime Minister Winston Churchill told the House of Commons, to its immense surprise, of the setting up of a Royal Commission to investigate how the laws on both mental illness and mental deficiency were working. While recent revelations of mistreatment and miscertification of the mentally ill had been causing alarm, it was undoubtedly the National Council for Civil Liberties' mental deficiency campaign that had prompted this decision. The Kathleen Rutty and Jean P— cases in particular had cut through; around forty questions had been tabled in parliament about the latter, and the 'liberty of the subject' issue now had new life breathed into it. It was a 180-degree turn for Churchill, who forty years earlier had pushed so hard for maximum restraint of the undesirables.

Sitting from February 1954 to April 1957, the Royal Commission on the Laws Relating to Mental Illness and Mental Deficiency took evidence from forty-two bodies but just eleven individuals. They rejected almost all personal correspondence on the topic of mental illness, and the unused testimony sits in the National Archives under the Ministry of Health category heading 'Rambling Letters'. (This file features all the usual Whitehall hard-heartedness and lack of interest in following up the complaints concerning brutality, theft and so on taking place in mental institutions. But that is another book altogether.)

With mental deficiency, many years had passed since any new legislation had been considered. The forty-year-long controversy concerning the 'moral imbecile' and 'feeble-minded' categories had never been tackled. It was upon this very topic that the Royal Commission received its greatest amount of evidence and viewpoints, with particular regard to the seventy per cent of 'defectives' who were described as being on the borderline of 'normality'. The consensus of the complainants was that people who didn't perform well at IQ tests, who had temporary psychological reasons to appear 'abnormal', who had had their education interrupted, who were emotionally immature, were currently detained in the mental health system – their reappraisals often no more than a formality.

The NCCL witnesses reminded the Commission that custodial care was only ever intended to be for those who had been abandoned; cruelly treated; subject to parental inadequacy or neglect; accused or found guilty of a criminal offence and found defective upon testing; incapable of benefiting from education; or requiring supervision after leaving school – plus the iniquitous clause singling out the mothers-to-be of illegitimate children who were receiving Poor Law Relief payments.

The main recommendation of the Royal Commission, when it published its report in 1957, was that flexibility should be introduced so that individuals weren't frozen into a particular category and their fate sealed. As knowledge about the mind grew, the aim should be to monitor patients to check the true nature of their mental problem, and to treat it accordingly, with institutionalisation only for the most severely ill or helpless or dangerous to the community. Care and oversight for those who needed it should be undertaken in the community, with mental health social workers involved closely with patients.

As for the psychopathic personality, which had risen in prominence in the 1950s thanks to bloody, headline-attracting crimes, the unresolved problem of intellect remained: 'the higher the IQ, the less reliable it is as a guide to character and capabilities', the Report's authors noted. These were 'pathological defects or abnormalities of personality, which result in behaviour which makes it most necessary . . . to provide them with special forms of help or treatment and in certain circumstances to subject them to special forms of control.'

The follow-up legislation defined the condition in these words:

psychopathic disorder . . . a persistent disorder or disability of the mind (whether or not including subnormality of intelligence) which results in abnormally aggressive or seriously irresponsible conduct on the part of the patient and requires or is susceptible to medical treatment . . . Nothing in this Section [of the Act] shall be construed as implying that a person may be dealt with under this Act as suffering from mental disorder . . . by reason only of promiscuity or other immoral conduct.

This latter phrase looks like a response to the 'locking up' of unmarried pregnant girls and anyone else considered to be socially or sexually 'deviant' or exhibiting sexually non-conforming behaviour.

Nevertheless, deep anxieties about female sexuality and fecundity remained. Dr Rees Thomas of the Board of Control said he was relaxed about the release of those previously certified as mentally deficient, but that 'the exception would be the promiscuous feeble-minded woman of the below average mentality. This is the only known group that I am uneasy about. Does the [Royal] Commission think that sexual promiscuity in a woman of below average mentality would constitute a ground for regarding her as a psychopath of the higher feeble-minded variety?'

With regard to mental deficiency, the Mental Health Act 1959 saw the repeal of the Mental Deficiency Act and the abolition of the 'moral imbecile' and 'feeble-minded' categories. The terms 'mental deficiency' and 'mental defect' were replaced with the actually no less grotesque 'mental subnormality' and 'mental handicap'.

The Board of Control was abolished and in its place, Mental Health Review Tribunals oversaw admission and discharge of both mentally ill and 'subnormal'/'handicapped' patients – with at least one non-medical person to be on the panel.

Most high-grade inmates detained under the Mental Deficiency Act were now classed as 'voluntary patients'. Thus, they were free to leave – so long as they were able-bodied; so long as they had somewhere to go. 'Community care' was the new policy thrust, for both the 'subnormal' and the mentally ill, with an emphasis on training and work, rehabilitation and occupation. The hostel system, occupation centres, child guidance clinics and the creation of 'therapeutic communities' were where hopes lay.

In the mid-1950s, some thirty per cent of people certified as mental defectives had been in detention between ten and twenty years, and fourteen per cent for between twenty and thirty years. Five per cent had been held for over thirty years. Tens of thousands of people would now require 'de-carceration' and the careful planning of their routes back into everyday, mainstream life.

In all the hearings of the Royal Commission and the discussions about the 1959 Mental Health Act, no reference was made to the original aim of deterring the breeding of undesirable people. The whole concept had died so hard, it didn't even need to be referenced. For now.

The suddenness of this policy reversal (from incarceration to liberation) was felt by everyone working in the sector. A hospital social worker told researcher Sheena Rolph, with regard to the mental deficiency patients,

> All of a sudden it seemed as though we got the directive, you know, the 1959 Act, you've got to do something, as though it had been a crime for these people to be there. I can still see the doctor sitting at his desk with piles of blue files, going through them, saying 'This one can go out, this one can't.'

She continued, 'The date 1959 was emblazoned on my memory as a crossroads . . . the exodus of those poor things.'

The opening of the doors was, naturally, welcomed by many who had been detained for years. 'Grace' told authors Maggie Potts and Rebecca Fido that

> it were some big man up from London that sent for me. Said that I shouldn't be in there. There were a lot of girls that shouldn't've been in. If you could get up at five o'clock in the morning, clean a boiler, you're not 'mental' if you can do that . . . Committee come and doctor come. I was dressed nice, put a nice dress on. Made cups of tea for 'em and they said I shouldn't be here. I was out before I knew where I were. I was happy as anything! I knew I'd do it!

Grace had been in Meanwood Park Colony for thirty-three years.

Mabel Cooper, who had been in St Lawrence's Mental Hospital for many years, told Professor Dorothy Atkinson of the Open University how difficult sudden freedom could be. Mabel was very glad that, in her locality, a community care system, a team, had been established; this team was formed to help ease long-stay mental deficiency patients into a flat of their own. Otherwise the process would have been 'very frightening. You might as well have dumped them out on the street.' Mabel made reference to the persistent anxiety that many discharged patients experienced about being re-called – the idea of 'them' saying, 'Right, you all have to go back in the hospital!' That was why, Mabel felt, regarding the institutions themselves, 'it's important that they knock them down, and then people like me and a lot more will know that won't happen.'

David Barron recalled the suddenness of the exits from Whixley. Suitcases began to be delivered to the colony, which had by then been rebranded a hospital. The onsite tailor became busy with measuring up the boys and men for new suits. A strong rumour did the rounds that many were to be transferred to hostels 'on licence . . . [which] meant we could be taken back to Whixley if we did any little thing wrong'. The dormitory gradually emptied, though it took a year for David to get his transfer. In the event, he was given just one day's notice. Eventually, he was given a full discharge following examination by two doctors and a psychiatrist – he was now thirty years old. His co-author wrote that David subsequently found himself many jobs, including cinema ushering, hotel work, portering and shop work, 'but time after time, malicious rumour that he had been "in a mental institution" caught up with him and he moved on, or he was moved on.' In the following years David tried to commit suicide on three occasions. Eventually, he was able to find stability, moving into sheltered housing in Greater Manchester in the mid-1970s. In 1981 he wrote: 'I've got to watch I don't slip back into institutional ways . . . I still cling to some routines.' Every morning he would precisely re-make the bed as he had had to on the ward – as though an inspection might take place at any time.

One released woman, who had done fifty years at Rampton, entering the State Institution at the age of eleven, was described by a nursing sister after her release as having 'had her life absolutely ruined . . . She couldn't understand that the cutlery wasn't counted, and that she didn't have to have a locked bedroom.'

***

In the first ten years of releases, the general trend was swiftly to move out those who had a proven ability to hold down a job, and who had a relative still living – or who had managed to maintain some kind of close connection with somebody in the outside world. For others, the old issue of retaining the best workers within the hospital still held them in place – all the more so when it was clear that the patient had no living relatives or friends who could support them outside. Still others had become used to their daily routines, and liked the friends they had made among other patients; the outside world from which they had been torn away as youngsters was not always regarded with longing.

Periods out on licence occasionally reminded some of the certified that they were an object of fear or ridicule, with teasing, bullying and humiliation from workmates, neighbours, even total strangers. A survey undertaken by the King's Fund in 1972 of 2,000 patients in nine mental hospitals reported that

> the popular press and general public often have a picture of patients being incarcerated in psychiatric hospitals against their will [but] nearly half would be sorry if they had to leave, and indications are that this figure would be considerably higher if more of the elderly patients were included. Indeed the proportion of patients who accept the hospital as their home where they can avoid the problems, personal or financial, of living outside its open gates is sometimes a serious matter for concern.

The medical superintendent of Starcross Hospital in Exeter compared the facilities in his institution favourably with the hostel system – the likely first stop after release. He said that a hospital that can offer playing fields, dances, television, outings and holidays, and which is set in lovely surroundings, had no substitute in a local authority hostel, possibly placed in a town centre.

From time to time, medical staff would warn of the problem of releasing a patient into 'unsuitable' family life. Interviewed by Sheena Rolph, charge nurse Derek Osborne said, regarding some of the patients he oversaw, 'We didn't want them to return home – that was often where the problems had started.' Osborne, though, also understood the desire for freedom among the majority of the men he cared for: 'The outside world was somewhere they wanted to get to – they thought that was Shangri-La, they thought that would be lovely. They wanted to get out, and a lot did eventually make it.'

Others still – an unknowable proportion – had actually become mentally ill within the institutional system, as mentioned in chapter 9. Although this is not a book about the experiences of the mentally ill,* the activities of the 'anti-psychiatry' movement of the 1960s, plus a number of separate campaigns regarding conditions and the questionable detention of psychiatric patients, became part of the push to speed up the release of people still held under the 1913 Act. Barbara Robb's landmark book, *Sans Everything: A Case to Answer*, published in 1967, triggered mass publicity and formal inquiries into the experiences of geriatric long-stay patients in mental hospitals. This elderly captive and neglected cohort will have included people who, decades earlier, had entered the system as 'moral defectives'. As the years passed, some will have gone on to develop either depressive or delusional illnesses.

This cannot be a surprise: they had been ripped from their families, or dumped into colonies by the very people who were supposed to love them, some with their new-born babies summarily removed from them, all without power or agency, and with no sense of ever winning their freedom. That is why we say girls were declared 'mad' for having a baby out of wedlock: it is true, but not in the way that is commonly believed.

***

There were other reasons why a patient originally admitted to a colony as a 'mental defective' could find themselves in a psychiatric institution. Capacity issues and purely administrative crises saw the shunting around of inmates from colonies into mental hospitals. At Lancaster Moor Hospital, the 1959 Act led to a huge reclassification programme of the hundreds of patients detained there. The medical superintendent found that twenty per cent of the females and fifteen per cent of the males were not mental patients proper: they were from the mental deficiency system

---

* Since this book is not about mental health itself, the history of Care in the Community for those with mental illnesses falls outside its scope. The best work on Britain's recent mental health policies and their implementation are: Richard Moth, *Understanding Mental Distress* (2022); Charles Brooker, *Serious Mental Health Problems in the Community* (1998); and Helen Spandler, *Madness, Distress and the Politics of Disablement* (2015).

but had been transferred across from what the superintendent described as 'the Merseyside conurbation' during wartime evacuation. The Second World War had created massive upheavals of this kind across the nation; and once better times (and a progressive government) arrived, matters did not automatically return to the *status quo ante*. Many patients remained displaced persons within the British mental healthcare system, and the ties to those on the outside who once may have taken an interest in their situation became yet more attenuated.

The reclassification and case assessments that were going on after 1959 sometimes compelled senior staff to make difficult decisions with regard to those detained originally as 'moral defectives'. The medical superintendent at Calderstones Hospital (Calderstones Colony as was) in Lancashire described, in 1963, his dilemma regarding two male sex offenders whose cases came up before the new Mental Health Review Tribunal. The first was a thirty-year-old man who had been certified as a moral defective in 1947 after three indecent assaults upon teenage girls; he was described as having a mental age of twelve-and-a-half. In 1954 he was discharged from his certification, and two years later indecently assaulted a woman, and was recertified. His assaults had all involved 'touching the privates' of the victims, as the Calderstones medical superintendent put it. The superintendent wrote that the patient looked 'normal', and was clean, tidy and fully co-operative in hospital. His parents were loving and very attentive, and his employment prospects were excellent since his father was a foreman at a building company and could easily find him work there. 'After thirteen years in hospital, successful holiday leaves, and parents with good insight into the patient's previous behaviour problems, was there justification in detaining the patient further?', the superintendent wrote. 'It is impossible, however, to state that any patient will not repeat antisocial sexual behaviour.'

With his second case, another man, also thirty, and with a mental age of ten, had been certified as a moral defective in 1949 because of his history of indecent touching of the genitals of three girls, aged five, seven and seven; also of breaking and entering; and of stealing a sum of £180. In 1953, he was allowed home from Calderstones on licence, but after five months the licence was revoked when he sexually assaulted a nine-year-old girl – again the act consisted of 'feeling the privates'. He, too, had a supportive family, decent work prospects and was 'normal'-looking, clean, tidy and co-operative. As *The Lancet* put it:

> After more than twelve years in hospital, where his behaviour has been good, was there justification in further detaining this patient? Further 'medical treatment' (as defined in the Mental Health Act 1959) was thought not likely to render this patient more fit to return to society.

The Calderstones superintendent was acutely aware that the criminal track record of these men was highly worrying; and yet he could not see indefinite detention as proportionate or fair. The tribunal agreed, 'after a long and careful evaluation', and both men were freed into the care of their families. The superintendent was anxious about either of the likely outcomes – indefinite detention; or the possibility of further sexual offending by the patients. The 1959 Act had empowered families and friends to be able to fight for the release of a patient, via the tribunal system. In terms of the 'liberty of the subject', this was an advance. But as the superintendent pointed out, 'a relative cannot be assumed to have a true insight into the behavioural problems of the patient'.

***

The stories of released people who had originally gone into the system as mentally defective made fantastic fodder for the tabloids, across the 1960s and 1970s. 'Sane – But Locked Up for 17 Years: The Girl Who "Died" Starts Life Again' was the *Sunday Pictorial*'s take on twenty-seven-year-old Peggy Richards – another triumph by the National Council for Civil Liberties, in 1961. Richards was released from 'hell' in Rampton following a Mental Health Review Tribunal pushed for by the NCCL. Peggy was the first patient to be freed under the new tribunal system (there were eighty-five people awaiting a tribunal hearing at that point, twelve of whom were in Rampton). Peggy had been removed from her home at the age of eight for being mischievous, high-spirited, a bit light-fingered: 'She often jumped on other girls' bicycles and pedalled off laughing. I was always having to give her a good hiding', Peggy's elderly mother ruefully told the *Pictorial*. A local authority official told Peggy's mother and father that Peggy should go into a 'foster home' for two years where she could be trained to be 'a good girl'; they had no idea what they were agreeing to. In fact Peggy went into the first of a series of placements and at the age of twenty-one ended up at Rampton. The week of her release, the *Pictorial* photographed her on a shopping trip and reported her real-time, verbatim comments as she was driven into the Cornish village where she had grown up and was reunited with her aged mother in their tiny picturesque cottage.

Cases that came to light more slowly included the three elderly women found during a mass reclassification exercise at St Catherine's Hospital for the Mentally Handicapped near Doncaster in May 1972 – twelve years after the legislation that permitted the institution to review all its patients. Two of the women were aged sixty-eight and another seventy-four, and each had been 'put away' in adolescence in the 1920s for having a baby out of wedlock, under the 'moral imbecile' clause of the 1913 Act. They were each considered sane but 'institutionalised'; all three had lost contact with anyone on the outside who once knew them.

One of the sixty-eight-year-olds had been certified after being impregnated at nineteen; her widowed mother, who was in poor health, had signed the paperwork that the authorities placed in front of her, without questioning its meaning. 'All I want now is a little job doing domestic work', the woman told the *Daily Mirror*, contemplating her release. The other two women were released into an old people's home in Dewsbury.

The Doncaster discoveries prompted other health authorities to do a similar trawl of long-stay patients, to see whether they included Mental Deficiency Act admissions who could now be freed and helped to establish some kind of independent living. It begs the question why this kind of 'census' exercise wasn't undertaken automatically and thoroughly by all regional hospital boards the first time around – as soon as the 1959 Act came into law. For what it's worth, the National Association for Mental Health (later renamed MIND) extrapolated from the Doncaster scandal that there were likely to be 30,000 people still being detained for 'social reasons' in mental hospitals.

It is likely that the inadequacy of money, staffing levels and capacity was, once again, a main cause of this ongoing detention of the formerly undesirable. The staff and facilities that could have made community care a success after 1959 were all too often missing in many regions. A common theme in this book is the failure of elaborately constructed policies because money and organisation were always lacking to put them into full and efficient operation.

By the early 1970s, localised community care options were still not functioning properly. 'Janet', thirteen years old, was discovered on an adult ward of Stanley Royd Mental Hospital in Wakefield – because there was simply no other provision for her in her part of the country. Illegitimate, with failures in her foster placements, and performing badly at school, despite an above-average IQ, Janet joined a gang of children in burning down a school outbuilding. She needed to be placed somewhere – but the only slack in the system locally was in among adult psychiatric patients.

***

Disdain – eugenically tinged – for the underprivileged, for socially outcast girls and women in particular, of course continued into the final quarter-century and on into our own era. In October 1974, Sir Keith Joseph, Conservative shadow Home Secretary, rehashed many of the attitudes and tropes from the early-twentieth-century debates on birth rates and social class. He told a meeting in Edgbaston, Birmingham, that 'the nation was moving towards degeneration'. He said that the use of birth control should be promoted more actively among social classes 4 and 5 (the low-skilled and the unskilled) in order to 'remoralise' society. 'The balance of our population, our human stock, is being threatened', he announced, because the following types of women were having children: those with a low IQ, the unmarried, divorced, deserted. 'They are producing problem children, the future unmarried mothers, delinquents, denizens of our borstals, sub-normal educational establishments, prisons, and hostels for drifters.' (Note that Sir Keith here revives another lovely old tradition in assuming that women were self-impregnating – the fathers escaped all notice, let alone censure, in his thinking.) 'A high and rising proportion of children are being born to mothers least fitted to bring children into the world . . . Some are of low intelligence, most of low educational attainment. They are unlikely to be able to give children the stable emotional background, the consistent combination of love and firmness.'

Labour secretary of state for social services Barbara Castle responded much as the early-twentieth-century environmentalist thinkers had. Society, she said, remoralised itself through keeping families and communities strong by eradicating poverty. Calling Sir Keith's speech 'frighteningly irresponsible', she argued that the development of social provision over the preceding half-century had brought about greater family, and social, cohesion. 'It is dangerous nonsense to suggest that the poor are parasites of welfarism.' And in any case, Sir Keith had shot his own fox since he intended to levy prescription charges on family planning products – hardly the way to encourage their usage. (The Family Planning Association, for its part, denounced Sir Keith's call for 'selective birth control'.)

It is interesting that his speech caused a national furore: castration, gas chambers, stud farms, thought police, 'incipient social fascism' – all these featured in some of the denunciations he received. Also interesting is that he wrote a long letter to *The Times* – amazed that he had hit a nerve, and while not exactly backtracking, he attempted to set his ideas in the context of a larger range of planned social services help for one-parent families.

The Birmingham speech was viewed in Westminster circles as the first shot fired in a Conservative Party leadership bid against Edward Heath. It can also be considered as one of the early attempts to widen the fissures between the Conservative right and the Conservative 'Wets': a culture war, if you like. Some within the party were very unhappy at Joseph's outburst; others were glad to see an energetic new rightwards drift. In the event, Joseph would throw his weight behind Margaret Thatcher in her leadership challenge. She replaced Heath in February 1975. Joseph would be her closest political ally, and they spoke as one in their hostility to welfare and towards the sections of the working class who appeared not to be 'aspirational' for status and possessions.

Thatcher's 'greatest achievement' was New Labour and the Blair government; and in 2006 a policy that would be satirically dubbed 'FASBO' (a foetal anti-social behaviour order) was introduced. 'Unborn Babies Targeted in Crackdown on Criminality: Blair launches policy imported from US to intervene during pregnancy to head off antisocial behaviour' was the headline in the *Guardian*. First-time mothers under the age of twenty-three from 'deprived communities' were to be identified between the sixteenth and twentieth week of pregnancy and targeted by nurse practitioners with a range of physical and psychological interventions to make sure their child did not become an enemy of civilised values. The scheme, the 'Nurse Family Partnership' (NFP), is voluntary and the intervention lasts until the child is two. Developed at the University of Colorado, the programme was introduced into Britain by Blair (and retained by subsequent administrations) in his attempt to 'tackle' the allegedly 'problematic' two to three per cent of British families: 'Some of these families actually cause wider social harms. The community in which they live suffers the consequences', Mr Blair said, with a straight face.

Tucked away in the Department of Health's explanation of the Partnership's objectives are 'fewer subsequent pregnancies and greater intervals between births'. Has this a flavour of eugenics? It's certainly fatalistic and it is class-based – predicated on the likelihood that a 'type' of woman is going to produce a problem child, and to have too many of them.

A survey undertaken by *The Lancet* in 2015 showed that in Britain, the Partnership had made no impact at all. Regarding its randomised trial of 1,645 interventions, *The Lancet* found 'no evidence of benefit from NFP for smoking cessation, birthweight, rates of second pregnancies, and emergency hospital visits for the child.' Nor did the young women go on to have fewer children.

Party political people tend to believe that class war pronouncements are effective in garnering votes. The headlines write themselves when whipping up hatred for societal 'scum'. As I write, the two main parties in England are placing the punishment of anti-social youths at or near the top of their offer to the electorate. One hundred and forty years on, and the same viewpoints and the same tropes are doing the rounds. Will they, will we, ever stop obsessing over the breeding patterns of the poorest and most outcast in our society? I want to be hopeful. I'm not hopeful. I'll try to be hopeful.

19

FORTY-SIX YEARS OF WASTE

Thousands of lives were wasted in the mental deficiency system in the forty-six years of the Act's operation – a colossal squandering of human potential. Many of those who were detained under the Mental Deficiency Act went to their graves without their stories ever being heard – the bulk of them, in fact: I have only been able to scrabble together a few of these case histories. The rest is silence.

There has never been a reckoning – as there has been, for example, with the 100,000 or so girls and infants who passed through the Irish Magdalene Laundries and 'mother and baby homes'. No public inquiry, no apologies, no curiosity, no acknowledgement that an untested hypothesis – that criminal and anti-social behaviour were biologically transmitted – was made use of to drive through this draconian legislation. Eugenics was a handy tool with which to corral and render powerless people who were seen as a hindrance to the challenges facing twentieth-century Britain. When the tide went out on hereditarianism, the law nevertheless remained in place. The focus turned to the prevention of bad citizenship and bad parenting, and keeping the labour market safe from inefficient workers.

But we cannot let the British public off the hook: yes, this was a brutal, blunt-force way to neutralise people who didn't fit in, or who were considered a burden in some way, but it could only have been carried on because of the passive acceptance of the populace at large. While many parents and relatives were, we now know, tricked into signing forms for colony detention, the majority handed over their troublesome youngsters without pressure being applied to them. And as the brief accounts in Appendix 2 suggest, many families chose not to visit or keep in touch, once their family member went into the system.

Policymakers were acutely aware of public opinion – and, after 1928, the opinions of a fully enfranchised population increasingly needed to be taken into account. We know that plans to ban 'defectives' from marrying were shelved in 1912 because of their likely unpopularity with the public. We know that even voluntary sterilisation was eventually given up as an impossible measure in the 1930s. But the public could, and did, tolerate the unnecessary detention of thousands of people on purely social grounds.

George Scott-Rimington told the National Council for Civil Liberties that the 'general public are too apathetic about these things'. This is a bit of an exaggeration, and 'ambivalent' rather than 'apathetic' is a fairer assessment. As we have seen, the public was easily aroused to anger whenever a newspaper decided to champion someone trapped in the mental deficiency system. Yet thousands also saw the appeal of dumping their exasperating, embarrassing or deadweight family member into a colony. When, in the 1950s, civil rights activists, together with psychologists, social workers and educationalists, pressed hard to overturn the mental health and mental disability system, they were not responding to a massive grassroots pressure group.

Historians have disagreed in recent years about the extent to which mental health and mental disability legislation can be said to be part of a 'social control model'; it has been argued that such a view fails to take account of the complexity of power relations, and the 'agency' of the patient. But immersion in the available case histories of those certified as moral imbeciles makes it hard to see it as anything other than a method of controlling certain sectors of society. Britain did decide to dispose wholesale of people whom it would not take the time, or spend the money, to help. Such time-consuming and costly interventions for the so-called 'socially inadequate' were abandoned for the lock-and-key of the colony. It would be perverse, for instance, to see the mass incarceration of girls and women who had illegitimate children as anything other than a patriarchal control system; similarly, the vast majority of people 'put away' were of the working class, or sometimes lower-middle class – and overlooking that power imbalance seems naive too.

Collectively, we have found a way of letting our nation off the hook for this dreadful chapter in our recent history. We tell ourselves that these were women and men who had been, in a former benighted era, incarcerated in *lunatic* asylums for not obeying society's rules – that they had been 'binned' for being 'mad'. More particularly, this is how the potent myth that 'the Victorians locked up girls for getting pregnant' has its genesis. But as this book has shown, this was a twentieth-century horror show.

AFTERWORD

THE WAY WE LIVE NOW – DEPRIVATIONS OF LIBERTY

As this book goes to press, over two thousand people with learning difficulties and/or autism are locked away, against the wishes of their parents or carers, and certainly against their own wishes. They are being detained in Assessment and Treatment Units, which are the successors to the long-stay hospitals. Two-thirds of the detained have autism. The average length of stay is five years and three months; some 350 have been detained for over a decade.

Yes, this is a swerve – *The Undesirables* is a book about individuals shut away for social or 'moral' reasons, rather than because they have an agreed, evidenced neurodevelopmental disability or difference. Nevertheless, the detention of people under mental health legislation who have no need to be detained reveals parallels between our era and the twentieth century.

People are being incarcerated unnecessarily because – yet again – this country will not spend the money that is required to supply the most humane forms of care; and the official mind retains its tendency towards inflexibility – unwilling to include families in decision making or to promote community care in small, local settings. Another continuity is the difficulty in finding staff – which is often, also, linked to poor funding. Why would anyone take such a demanding job if the pay is poor? Even in the supposed 'golden age' – the postwar years on to which we project the notion that government was committed to providing high-quality public services – the archives tell us that mental health and mental disability remained the 'Cinderella services', denied the money and attention they needed to function as planned. Never enough staff; never enough beds; never enough community care provision.

In our era of ongoing austerity, central government has been underfunding local authorities for many years, severely impacting on their ability to support people living at home with their own families, or in assisted-living facilities. Places in secure units and Assessment and Treatment Units, however, are funded by the NHS; this means that decisions are all too often being taken in order to make a saving to the local authority budget. What's more, autistic people can be 'sectioned' under the 1983 Mental Health Act, which is ironic, since that piece of legislation is best known for substituting 'care in the community' for hospitalisation.

'Mechanical restraint', meanwhile, was outlawed in the late 1830s within the English lunatic asylum system. Today, handcuffs, belts and harnesses are used to restrict the movement of inmates. Seclusion, too, was controversial under Victorian lunacy law – we now have reports of people spending days or even weeks locked up entirely alone. This is to calm them down, we are told; and it prevents them from harming themselves or others. But seclusion is also convenient for facility managers when staffing levels are low. Five-and-a-half thousand incidents of mechanical or chemical restraint or seclusion were reported in the month of January 2023 alone, according to NHS data.

***

At Christmas 2009, Mark Neary went down with a serious bout of flu. This left him temporarily unable to care for his twenty-year-old son Steven at their home. Mark arranged three days of respite care with the London Borough of Hillingdon – at a respite home Steven had visited before over the previous two years, without problem. Steven has learning difficulties and autism. The respite centre to which he was sent transferred Steven to a unit for assessment, alleging that he had behaved aggressively while at the centre. Institutional detention has a strong tendency to exacerbate the very problems that the patient is experiencing, particularly in the case of autism. It is the opposite of curative. As is the case for most people with autism, routine is crucial to Steven, and he did lash out at the respite centre. But many months later, his assessment had not been carried out – he was still in the unit, to his distress and to the distress of his father.

The Council then put forward the argument that Steven was not a suitable person to live at home with his father – odd, because he had had no problems at home before. They gave no reason for this decision, and both Mark and his son were kept entirely in the dark about the process. Steven consistently expressed a desire to return home; Mark consistently expressed the desire to have him home.

Mark pursued his case, and a landmark court judgment in June 2011 found that Steven's rights to a family life and to liberty under Articles 5 and 8 respectively of the Human Rights Act had been breached. Steven was freed.

This ought to have set the pace for subsequent procedure. But sadly not. Alison Rodgers was told that her son Adam, who is autistic, would be cared for for six months at a treatment centre; fifteen years later, he is stuck in Rampton – his entire adult life having been spent in secure units. It is exactly this type of environment that makes Adam's condition worse. His fellow inmates include Ian Huntley, the 'Crossbow Cannibal' (Stephen Griffiths) and armed robber 'Charles Bronson' (Michael Peterson). It costs a quarter of a million pounds a year to keep Adam in such company, but if he were at home with Alison, it would be a fraction of the cost. Alison, distraught, continues to campaign for his release.

Over the years, there have been any number of superb exposés of unnecessary detentions of this kind, by both broadcast and print media. Nothing is hidden from us anymore. These incidents make headline news. But while there is no shortage of sunlight, it is no longer acting as a disinfectant. Beatings, taunting, drugging, solitary confinement – all are captured in footage and seen by the public and parliamentarians. Then there's an official condemnation, then a review, then some pledges. And then it happens again.

As I write, another attempt at fine-tuning the status of personal liberty is under way. The draft Mental Health Bill is undergoing parliamentary scrutiny, with the hope of un-yoking autism and learning disabilities from 'mental disorder'; as well as imposing stronger requirements on local authorities and the NHS to plan for patients' discharge. This all sounds good, but how are they supposed to meet these new requirements if they are not given money to do so?

The Mental Capacity Act 2005 has also very recently been tweaked. The 2005 Act introduced Deprivation of Liberty Safeguards, operational from 2009. The Act was designed to protect and empower people who may lack the mental capacity to make their own decisions about their care and treatment. The Court of Protection was set up to oversee the process.

The Safeguards were a set of criteria that had to be met to ensure that all restrictions of personal liberty were reasonable and proportionate. The people for whom they were devised were, typically, the elderly with dementia – whether in residential care, in hospital or in their own home; someone who has sustained a brain injury following an accident or a serious illness; as well as those with autism or a learning difficulty.

Officially, a person is deprived of their liberty if s/he is 'under continuous supervision and control and is not free to leave, and the person lacks capacity to consent to these arrangements'. For a deprivation of liberty to be lawful, the care must be 'in the person's best interests', and any restrictions must be necessary and proportionate.

Orders are imposed to restrict a resident's right to leave a care home, assessment centre or hospital, or even their own residence if care is being provided at home.

Making a decision on behalf of someone who lacks capacity is supposed to involve making a choice that maintains as far as possible their basic rights and freedoms, or the 'least restrictive alternative'. In practice, however, local authorities are overwhelmed by the number of applications they receive to authorise deprivation of liberty, and a large backlog of over 100,000 unprocessed applications has built up. For this reason, some people are being deprived of their liberty during the processing of the application. The Joint Committee on Human Rights puts it this way: local authorities are left having to work out 'how best to break the law', because they haven't got the resources to process this many applications.

The Court of Protection itself came under fire almost from its inception for being slow, incompetent, overly bureaucratic and expensive for the public to deal with. As with the Board of Control, it was criticised for a 'behind closed doors' way of operating, with levels of furtiveness out of all proportion to the need to ensure privacy in sensitive family matters. However, the court has in recent years introduced a streamlined process to deal with its own backlog and most of the problems are, as mentioned, further back in the pipeline, with the underfunded local authorities. The

2005 Mental Capacity Act has been modified (or, rather, had its operations convoluted) by various amending Acts. In the opinion of one lawyer, 'I have no doubt that this area of the law is so complex that cases ought to be handled only by absolute specialists.'

More fundamentally, as mental capacity law expert Lucy Series has written, 'The deprivation of liberty safeguards are so complicated that even senior judges have described them as like "putting your head inside a washing machine and spin dryer". It can be very difficult for people with disabilities, older people, and their families, to use them to secure their freedom.'

Cases that have been challenged in the Court of Protection reveal the level of over-reach by those imposing the orders, using a mask of compassion and concern to interfere in the lives of people who should be free to make their own decisions – even if these turn out to be unwise. 'What good is it making someone safer if it merely makes them miserable?' said High Court judge Sir James Munby in 2007.

Modern deprivation of liberty is defended in the name of protecting the vulnerable from mishap if they are out and about; from falling and injuring themselves in their own home; and from causing such incidents as fires and gas leaks, for example, through forgetfulness. But there is a body of evidence that liberty is being denied because it is a convenience for care homes that are poorly staffed, or because families no longer have the time or the inclination to oversee elderly relatives living in their own homes. There may also be fears of litigation arising from the actions of those who have mental capacity problems – fears that they might wreak havoc if they were at liberty to cause accidents, for example. And no one (whether family member, social worker, or care-home manager) wants to face blame if someone has indeed been left free to be injured or killed because of their vulnerability. So we also prevent them being a risk to themselves by limiting their movement.

But isn't it also to do with our not having the patience or imagination to accommodate people who are no longer going at our speed, sharing our interests, thinking and behaving as we do? Do we tidy them away when they become inconvenient to us? Is there a cultural problem, that we see unusual, unorthodox, seemingly inexplicable behaviour as something that should be legally constrained – placed out of sight, tucked away?

A replacement system, Liberty Protection Safeguards, was planned for introduction in 2021. Cynics viewed this as no more than a rearrangement of the words in the name. The aim was to make it easier for family and friends to be involved in the care of the person who is incapacitated, to tighten criteria for the capacity assessment itself and to 'place the person at the heart of decision-making and [to be] compliant with Articles 5 and 8 of the European Convention on Human Rights.' However, the Liberty Protection Safeguards system has been repeatedly delayed by government, and some experts in the sector doubt whether it will ever be implemented. They fear that governments in general are unwilling to grapple with the complexity and cost of this issue.

We have some useful blueprints for a properly functioning mental disability – and mental health – system. It'll take good and consistent funding; decently paid, carefully trained and monitored staff at every level, including – especially – at the top; fantastic national and local organisation and administration.

And the Court of Protection does seem to be more inclined in recent months to see things from the point of view of the person fighting for their liberty (and many case hearings are now even streamed online). In a case where an elderly man was being detained in a care home, against his and his family's wishes, Mrs Justice Lieven told a local authority, in the summer of 2021, 'You can't just exist in this extraordinary world where he's being deprived of his liberty but no one seems to care very much. Just because he's old and infirm does not mean you can deprive him of his liberty! You have to do better than that. You can't ask the High Court to turn a blind eye to illegal detention . . . You can't unlawfully detain people in the UK.'

APPENDIX 1

HOW THE 1913 MENTAL DEFICIENCY ACT WORKED

Parents (or a relative acting *in loco parentis*) could 'petition' the Mental Deficiency Committee of their local authority to have an alleged defective under the age of twenty-one examined, with a view to having them certified as mentally deficient, and stating that they themselves were unable to provide the care and training that their child required.

If any private individual other than a parent or relative were the petitioner, the consent of the former must be sought, unless 'such consent is unreasonably withheld'.

Certificates were to be signed by two qualified doctors, one of whom had to be expressly approved for such examinations by the local Mental Deficiency Committee. These two certificates were to be accompanied by a statement – a statutory declaration – signed by the parent or relative; the declaration would state the reasons why the individual came under the definition of mentally deficient in the new Act, and which of the four categories of deficiency applied, and why.

If a suspected defective was found 'neglected, abandoned, or without visible means of support, or cruelly treated', the local Mental Deficiency Committee was obliged to arrange their assessment.

Additionally 'subject to be dealt with under the Act' were those found guilty of a criminal offence (but in advance of their sentencing) who appeared to be exhibiting signs of mental deficiency. If a young person was up before the magistrate on a charge, the JP could even order an assessment ahead of conviction if the offence and the defendant's behaviour suggested they were deficient.

Also in the frame for assessment were people already held in a prison, reformatory, industrial school, inebriate reformatory for habitual drunkards or an asylum for criminal lunatics.

Local education authorities (LEAs) were obliged to assess and report children over the age of seven 'who have been ascertained to be incapable by reason of mental defect of receiving benefit or further benefit in special schools or classes'. In addition, under a linked measure, the 1914 Elementary Education (Defective & Epileptic Children) Act, children in special schools approaching age sixteen were to be assessed by the local education authority with a view to being sent to an institution or placed under guardianship.

The 'named guardian' was to be sourced via the local authority or one of the voluntary associations that worked alongside the local authority; quite often, a guardian was a relative of the defective.

Upon an individual's arrival in an institution (usually a mental deficiency colony) or a guardian's domestic home, detailed notification was to be sent within seven days to the Board of Control – the new Whitehall body that took over from the Victorian Commissioners in Lunacy oversight of both the mentally ill and the mentally disabled. The Board itself could decide to intervene if an alleged defective had not been assessed despite there being reasonable cause to believe they were in need of certification. Critics would later say that the new Board of Control was a regressive step, because, unlike with the Commissioners in Lunacy, only the Board could consider the discharge of a patient; and henceforth, no family member, or friend of the patient, could obtain an independent assessment with a view to release.

If a parent or relative wished to retrieve their family member from a colony or guardianship, they were to contact the Board of Control and if the Board agreed that the family had the means to care for and supervise the defective, they could be permitted to live out on licence. If, however, the Board did not agree, the family could not apply again for another six months.

APPENDIX 2

ANECDOTES OF WOMEN DETAINED FOR HAVING A CHILD OUT OF WEDLOCK

In January 2022 photographer Ian Beesley tweeted one of the photographs he had taken during his time documenting the closure of Moor Park psychiatric hospital, in 1996. It features 'Dolly', an elderly lady who had spent her life in the institution, having originally been detained under the 1913 Mental Deficiency Act because she had had a baby at the age of fourteen.

Beesley's tweet attracted a huge response, with many replying with their stories of similar cases. Some respondents had trained and/or worked in the mental health sector; others knew fragments from their own family history. This is a selection of the replies, anonymised.

- 'In 1980 as a student nurse I met an old woman who'd been deemed "a moral defective" under the 1913 Mental Deficiency Act & committed for life simply for becoming pregnant while "in service" as a teenager. Of normal intelligence, but after 60 years completely institutionalised.'
- 'I met a woman like this too. I couldn't believe it when I read the notes on the reason for her being admitted. "Morally defective." She got out about 20 years later, as part of a wider community resettlement programme, but she was massively institutionalised.'
- 'Happened to one of my great aunts, she spent her whole life in Doxford Hall. Split the family too – my grandfather opposed sending her away.'

- 'I had a great aunt who exactly the same happened to. I only found out when researching the family tree. The sad thing is she only died in the 1980s. Family knew about her and didn't tell me.'
- 'I met a woman in one of those hospitals in the 1990s. She had been put there during WW2 because she kept "talking to soldiers". Incarcerated for being "morally degenerate". All she did was talk to soldiers who had been billeted nearby during the war. 50 years.'
- 'I met an elderly lady c.1982 when doing some voluntary work in a psychiatric hospital. Same story, she was 14/15, family were well heeled. She kept asking [me] to keep a look out for her parents who were coming to get her. Heart-breaking.'
- 'The large county psychiatric hospitals where I did my training had numerous elderly ladies who had been admitted as young girls for similar reasons. Most disturbingly, they had often reported sexual abuse by an adult.'
- 'This happened to my dad's mum and when he was 18 he tried to get her out but by then she was so institutionalised that she was distraught if she was left for more than a few hours.'
- 'My cousin suffered the same horror at York. She wrote pleading letters to my grandmother asking to come home. Devastating.'
- 'I used to work at Winwick Hospital, one of Europe's largest psychiatric institutions, back in the day. There were multiple stories like this one. Many had been there so long, they couldn't leave. They were co-dependent on the hospital & its staff.'
- 'I too worked in a psychiatric hospital. These women who'd had babies out of wedlock were put in there. Classed as moral defectives. Became so institutionalised, they would not cope outside.'
- 'I worked in elderly care for many years from 1991 onwards. Histories like this lady always got to me, their lives and children taken away by authorities. Heart-breaking.'
- 'Women like this were known as "moral defectives". The learning disability hospital I trained in had several women who had very mild LD [learning difficulty] but were originally admitted as moral defectives... they were all in their 80s+. Very sad.'
- 'Close by is a beautiful old building that used to be a psychiatric hospital (it's been luxury apartments for 18 yrs). 25 yrs ago when I moved to the

area I met many locals who were employed there. They told me there were lots of patients like Dolly. I felt shame then & still do.'
- 'My friend's mum worked in a similar hospital in the 70s and 80s and said exactly the same thing. Within a few years, patients became so institutionalised that they were unable to function in the outside world. The wards were full of very old ladies like this.'
- 'Back in the 80s I worked in a large psychiatric hospital in London and one of the patients there was a lady of 90 years old, who had had a child at 16, and was placed in the hospital. She never left either. Broke my heart.'
- 'I trained as a nurse in Edinburgh early 90s & met a lady who had been in long-term psychiatric care due to having a child at 16. Her story stayed with me. She said she was "sane" when she was admitted but became mentally ill due to horrific impact of institutionalised care.'
- 'I worked in an Essex psychiatric hospital in the 80s. I met many women like that. Locked away, declared mad and institutionalised for nothing more than having a baby. Heart-breaking.'
- 'In the 1970s when I was a teenager I used to volunteer to play cards and chat to patients in one of 5 psychiatric hospitals in Epsom. There I met women who had been incarcerated for decades for having children while under age. Many of the patients were kept heavily sedated.'
- 'My school sent us to do community service in the 80s in one of Epsom's many psychiatric institutions when we were 15. Never forget the elderly lady I met who held my hand & told me she'd been there since age of 15 for getting pregnant. Unbelievably sad!'
- 'I met a couple of older ladies who'd been locked up for this reason, during my nurse training in the late 80s. Old Victorian psych hospital. Their life stories are heart-breaking. Will there ever be justice for these women & their babies?'
- 'When I was a student nurse in the late 80s I was seconded to a psychiatric unit. The long-term ward had several elderly ladies who had babies in their teens, unmarried. I don't know what happened to the babies. I have never forgotten it. So cruel!'
- 'I trained as a mental health nurse back in the 80s and encountered several such cases. Heart-breaking.'
- 'As a student nurse in the 1980s I worked in a "psycho-geriatric" (1980s!!) ward. There was a lady there in her 70s incarcerated for the same reason.'

- 'This was commonplace in the 40s 50s & 60s, I worked in a rehab facility in the 80s where several women in their 60s lived who had been in Yorkshire psychiatric hospitals for 40 years, for being unmarried mothers.'
- 'When I was a sixth-form student in the 70s I did community work in a local hospital for people with learning difficulties. One older patient was an unmarried mother in the 1930s, two others were lesbians. The practice of lifelong incarceration was widespread.'
- 'I met an elderly lady with an identical story but she was released. Two things she commented on after her release, how kind people were to her, and how lovely it was not having to eat food with plastic utensils.'
- 'My mother was born in a psychiatric hospital. When I asked why, I was told my grandmother was angry! ... As a young speech & language therapist covering 2 institutions, I found this quite common.'
- 'My mother was told that her mother had died. Was a bit of a shock to see her death announcement when I was 14. My mother was so angry when she found out but it was all kept secret for her own good.'
- 'I knew a woman who was abandoned by her family and was placed in an "asylum" when, very young, she had an "illegitimate" child who was taken from her. She was finally discharged in her 70s and she tracked down her (still living) mother. Her mother refused to meet her. So sad ... She was completely institutionalised, childlike and vulnerable and was so excited when Salvation Army found her mum in a care home. She couldn't understand why her mum rejected her. Still blamed herself. This was 1980s and I still think of her.'
- 'I nursed a lady in the West Mids in 2002 who had a baby with an American soldier in WW2. She never left. She became mute, I imagine due to trauma.'
- 'When as a nurse I worked in both psych and general, there were loads of long-stay patients in there purely for having been unmarried mums. And of course they wouldn't have been able to function out of that environment. So sad.'
- 'My nan was placed in a psychiatric hosp early 1980s, early onset Alzheimer's. My uncle worked there & told me some of their life stories, met 1 or 2 ladies in same situation! None of them ever left.'
- 'In 1995 I went to a large hidden away home near Southport where twin sisters in their 80s were walking round from room to room to room, the

last of the patients living there. One of them had given birth when she was 13. They locked both of them away. Both of them . . . They wandered around all day. Holding each other's hand. I sat in the car staring out for 40 minutes before I could leave.'
- 'I remember whilst working in care, meeting women who were institutionalised for this reason. It's truly heart-breaking the effect it had.'
- 'My sister is a mental health nurse & when she was training in the 80s she worked in a large psychiatric hospital. She said there were people in there for being gay, for having kids out of wedlock, for refusing to get married or just for being a bit "odd". All put there by family.'
- 'Pregnant at 13 . . . she was being sexually abused. The trauma of that plus being locked away is just terrible. And the perpetrator probably got away.'
- 'They were assumed to be mentally ill because only a crazy person would have extramarital sex (I wish I was joking).'
- 'Often these girls were victims of incest. That is often why they were locked up – to hide someone else's shame and protect the family's reputation. It is probably also why so many never left. The complete betrayal broke them.'
- 'I worked in a hospital for acute mental episodes, also long-term patients, some of whom had simply given birth out of wedlock. One lovely lady was found in the attic of her mother's house. The mother died and this poor soul was secreted for years in the attic. She couldn't talk.'
- 'It takes two to tango. So what happened to the males involved? I guess they got away scot-free . . . There's no male equivalent of "slut".'
- 'Mostly, they would have been fathers, uncles, brothers or family friends of the girls who were "put away", so no consequences. In the 80s when working in the NHS, I met elderly men committed to asylums as children after being raped, or even for being caught masturbating.'
- 'There should be people/authorities held to account and compensation claims for the families affected. It's a scandal.'

APPENDIX 3

AMERICAN RESPONSES TO A BRITISH 1951 SURVEY ON STERILISATION

In 1951, eighty-eight questionnaires were sent out by the National Association for Mental Health (NAMH) Mental Deficiency Sub-Committee to American mental deficiency institutions, regarding their use of sterilisation. Thirty-five replies were received by the NAMH. Its chairman concluded that it was evident that very little detailed follow-up had been done in America on sterilised defectives and little consideration had been given to the issue by those who ran institutions.

In 1951, twenty-eight US states had sterilisation laws, of which twenty-four permitted compulsory sterilisation; only nine undertook sterilisation outside of institutions.

The total number of people in the US who had been sterilised since 1907 was 26,858, of whom fifty-one per cent were mental defectives.

These are the NAMH questions and replies.

'**Question 1:** do you find there is an appreciable number of patients for whom sterilisation is desirable before licence?'

Eighteen respondents answered yes.

Montana, State School: 'Only useful for a small group with a mental age of eight to twelve and lacking in moral concepts.'

Washington, Lakeland School: of 1,552 inmates, ten per cent were sterilised.

Idaho, Nampa State School: sterilisation is 'desirable for several hundred' patients; but relatives are often unwilling.

California, Pacific Colony: numbers are greatly decreasing. Now only an average of three are sterilised a year, compared to the sixty before the sterilisation law was made voluntary, in 1951.

North Carolina, Raleigh State Hospital: very few high-grade (borderline) patients are recommended for sterilisation, which is used 'conservatively'. About fifty of 2,400 patients have been sterilised.

Ohio, Columbus State School: sterilisation is not legally possible, so many who could be out on licence are detained in the institution.

Nebraska, Beatrice State Home: all patients must be sterilised before release. 'Enthusiastic claims are made for this practice.'

Connecticut, Mansfield Institute, Dr Neil Dayton: sterilisation is no answer to the problem of preventing mental deficiency with 'eighty-six per cent emanating from carrier group . . . Mental defectives are usually undersexed, only a few are interested in sex, very few marry and those who do, do not have large families . . . Therefore no point in sterilisation and Dayton does not use the law permitting it. In twelve years he has had only three pregnancies out of hundreds of girls placed out in the community, which is a much better showing than is made by the general population of this state in young women of the same age group.'

Vermont, Brandon State School: 'A considerable amount done in the past but recently there has been a sharp decrease and only one sterilisation reported in the state during the past year. This institution is not attempting to make sterilisation a part of its programme – always a subject for discussion.'

Massachusetts, Wrentham State School: 'There are many female patients in the institution who are capable or working in the community and we would be much more comfortable about releasing them on trial visit if they were sterilised.'

'**Question 2:** Have you found that sterilisation encourages promiscuous sex relationships, or exposes girls to undesirable attentions from the opposite sex?'

Montana: 'Doubtful if there is any difference, but sterility does attract certain men.'

Idaho: 'No effect one way or the other. All defectives are likely to become promiscuous in any case.'

California, De Witt Hospital: 'Relatives sometimes report that sterilisation exposes a girl to undesirable attentions from the opposite sex.'

Virginia, Petersburg State Colony: 'May encourage promiscuity in a few cases, but in general, no.'

Indiana, Fort Wayne State School: 'No. Sterilisation lessens this problem if the operation is performed several months before placement [in the community].'

Nebraska: 'We do not find that any of our girls who have been sterilised make any attempt to enter promiscuous sexual relations. In fact, we feel that our experience justifies the statement that there may be less interest.'

Connecticut: 'Has known one case where this has happened and considers that fear of pregnancy does not act as a deterrent and is therefore a valuable weapon.'

'Question 3 a): In the case of sterilised mental defectives who subsequently marry, have you found any resentment at the inability to become mothers, or that this inability sometimes produces neurosis?'

South Carolina: 'A few husbands have objected.'

Montana: 'In general, the approach is fatalistic but we have known husbands who have been indignant.'

Idaho: 'Yes, because every human being above the level of an idiot has the instinct to reproduce.'

California: 'Some ex-patients are reported to be unhappy and disappointed to a marked degree.'

Georgia, Training School: 'Very occasionally, request received by an ex-patient for another operation to permit child-bearing.'

Virginia, Lynchburg State Colony: 'In ten years, only two cases have shown any interest in the matter.'

Nebraska: 'In the case of our married sterilised defectives, we have had no resentment because of their inability to become mothers; or that this inability has at any time produced neurosis. We have been privileged to talk to a number of them and they have commented that they were very happy and are glad that they cannot bring children, who might be defective, into the world.'

Connecticut: 'Has known instances where sterilised girls have married later and have bitterly regretted their inability to have children. They and their families have criticised the school for having permitted sterilisation, even though at the time, consent was given.'

California, Sonoma State Home: 'Patients who subsequently married have shown considerable resentment – occasionally a precipitating factor in producing neurosis (refers to high-grade borderline only).'

'Question 3 b): Do many of your sterilised married defectives achieve a tolerably satisfactory home life and are they helped in this by the absence of the responsibilities of motherhood?'

Nebraska: 'It is our feeling that all of our people do have a perfectly satisfactory home life, and we know of several who own their own homes at the present time.'

Montana: 'Our records show a high degree of success with childless mental deficients who marry. In many instances, both parties work and attain a standard of living which satisfies them and does not interfere with anyone else.'

Oregon: 'Sterilised patients have been able to develop a perfectly satisfactory home life without the problem of caring for children.'

Acknowledgements

Thank you Sam Carter, Hannah Haseloff, Margot Weale, Laura McFarlane and the team at Oneworld; and my agent Sarah Chalfant at the Wylie Agency. Claire Weatherall at the Hull History Centre allowed me access to the National Council for Civil Liberties archive. Much of this book was researched between lockdowns, and I am even more grateful than usual to the staff at the National Archives, London Metropolitan Archives, the Wellcome Collection's archives and the British Library for their professionalism during those terrible months.

Peter Neish read the first draft – brilliantly, as he always does. Pauline Black of Manchester Metropolitan University supplied me with helpful source material for today's Deprivation of Liberty laws. And thank you Julie Matthias, Debbie Millett, Diana Maltz, Liz Tames and Karen Shook for letting me bend your ear about this grim project. Kathleen McCully did a superb copy-editing job.

I am a Victorianist making her first foray into more modern times. Among those who have laid the groundwork in the story of 'feeble-mindedness', 'social Darwinism' and twentieth-century mental health legislation, I am particularly indebted to: Mathew Thomson, Pauline Mazumdar, Clive Unsworth, Mark Jackson, Maggie Potts, Rebecca Fido, John Owen, Stephen Watson, John Macnicol, Greta Jones, G.R. Searle, Sheena Rolph, Dorothy Atkinson, Gillian Sutherland, David Wright, Anne Digby, Janet Saunders, Jan Walmsley and Phil Fennell. I have found their work invaluable in attempting to piece together the timeline and conflicted motivations that brought eugenical thinking into British social policy with the passing of the 1913 Mental Deficiency Act.

A massive thank you to Dr Lucy Series of the University of Bristol for her generous close-reading of my Afterword, and for helping me to understand the unholy mess we today find ourselves in with regard to mental capacity and personal liberty.

Celia Kitzinger and Gill Loomes-Quinn launched their wonderful website Open Justice: Court of Protection Project, reporting from the Court on the many issues regarding personal liberty today. It's a fascinating window on how our rights are negotiated, and it can be accessed here: openjusticecourtofprotection.org.

Notes

Unkind Words: A Note on Terminology
p. xi *Persons of unsound mind* . . . Report of the Royal Commission on the Care and Control of the Feeble-Minded, 323–5.

Foreword
p. xvii *The cases of detention* . . . National Council for Civil Liberties, *50,000 Outside the Law*, Foreword.

p. xvii *The rat-catcher* . . . Potts and Fido, *'A Fit Person to be Removed'*, 20.

p. xviii *Their propagation must be prevented* . . . A.F. Tredgold, 'National Association for the Feebleminded', *British Medical Journal*, 22 May 1909, quoted in Jones, *Social Hygiene*, 31.

p. xviii *in 1950 two psychologists random-sampled one hundred patients* . . . memo on mental deficiency statistics, National Archives, MH 121/44.

Chapter 1
p. 3 *Even when their houses are whitewashed* . . . Alfred Marshall, 'The Housing of the London Poor', *Contemporary Review*, February 1884.

p. 3 *The environment is the product of the individual* . . . George Mudge, 'Biological Iconoclasm, Mendelian Inheritance and Human Society', *The Mendel Journal*, October 1909, 49–50.

p. 4 *So long as we leave several hundreds of thousands* . . . Clarence Rook, 'St Patrick Hooligan', *The Anglo-Saxon Review*, December 1900.

p. 4 *The excitement of a town life* . . . Helen Bosanquet, 'The Children of Working London', in B. Bosanquet (ed.), *Aspects of the Social Problem*, 36–8.

p. 5 *Hereditary predisposition is a prominent cause of mental derangement* . . . Burrows, *Commentaries*, 101.

p. 6 *Moral insanity . . . is a morbid perversion* . . . Prichard, *Treatise*, 4.

p. 7 *Tuke's theory of 'dissolution'* . . . Tuke, *Dictionary*, 331.

p. 8 *Abortive beings in nature* . . . Maudsley, *Responsibility*, 209.

p. 8 *Maudsley . . . came to believe that . . . 'moral insanity' was an illness that arose from a congenital defect* . . . Walker and McCabe, *Crime and Insanity*, 210.

p. 8 *The conditions under which men of high type are produced* . . . Galton, *Inquiries*, 45.

p. 9 *What Nature does blindly, slowly and ruthlessly* . . . Galton, 'Eugenics', 42.

Chapter 2

p. 10 *In 1881 Dr William Guy . . . recommended lifelong detention* . . . Guy's memorandum 'The "Insane" and the "Imbecile"', *Report of the Commission on Criminal Lunacy*, 161–4. I am grateful to Janet Saunders for this reference, in her 'Quarantining the Weak Minded'.

p. 10 *Guy's proposal rejected as illiberal* . . . *Report of the Commission on Criminal Lunacy*, 164–7.

p. 10 *You see, these people I speak of* . . . *Report of the Royal Commission on the Penal Servitude Acts*, 1016, quoted in Saunders, 'Quarantining the Weak Minded', 278.

p. 10 *There is a great deal of nonsense talked about the liberty of the subject* . . . *Report of the Royal Commission on the Penal Servitude Acts*, 1017.

p. 14 *Within the prison system, from 1876* . . . Watson, *Moral Imbecile*, 152.

p. 14 *Clever fools* . . . Mercier quoted in Tredgold, *Note*, 7.

p. 15 *Many feeble-minded persons, owing to their defect* . . . Allan Warner, letter to the *British Medical Journal*, 19 February 1927.

p. 16 *Estimates put the 'weak-minded/feeble-minded' at around three per cent* . . . evidence of Prison Commission medical officer Dr Herbert Smalley at the Royal Commission on the Care and Control of the Feeble-Minded ('The Radnor Commission'), *Minutes of Evidence*, vol. 1, 175.

p. 16 *An organic anomaly* . . . Ellis, *Criminal*, 38, paraphrasing Cesare Lombroso's theory in Lombroso's *L'Uomo Delinquente* (1876).

p. 17 *Prison system was 'like a sewer'* . . . Ellis, *Criminal*, 98.

p. 17 *It must be remembered* . . . ibid., 21.

Chapter 3

p. 20 *Education on an empty stomach is a waste of money* . . . Jowett speaking in 1904 to Bradford City Council, Spartacus Educational (online).

p. 20 *We see numbers of half-imbecile children* . . . Tabor, 'Elementary Education', 218.

p. 22 *One per cent of children at elementary schools in England and Wales could be described as 'mentally deficient'* . . . evidence of Philip Bagenal, Radnor Commission, *Minutes of Evidence*, vol. 1, 132.

p. 23 *Living in very bad homes* . . . evidence of Arthur Acland Allan, chair of the LCC Special Schools Committee, ibid., 416.

p. 23 *To be found in the streets* . . . Mr A. Allen MP, *Hansard, House of Commons Debates*, Fifth Series, vol. 41, 19 July 1912, col. 735.

p. 23 *Their feeble-mindedness [shows]* . . . *more in their moral qualities* . . . evidence of May Dickinson Berry, assistant medical officer LCC (education), Radnor Commission, *Minutes of Evidence*, vol. 1, 544.

p. 24 *The LCC issued its own printed forms* . . . LCC forms and reports on schoolchildren examined for admission to special schools, London Metropolitan Archives, LCC/PH/MENT/2/002, Box 2.

p. 24 *What is the appearance of the child* . . . ibid.

p. 25 *Is the child of solitary habits?* . . . London Metropolitan Archives, LCC/PH/MENT/1/3.

p. 26 *There is an unwillingness on the part of teachers to report these cases* . . . evidence of Dr J. Mitchell Wilson, Radnor Commission, *Minutes of Evidence*, vol. 2, 81.

p. 26 *Statistical evidence is very incomplete* . . . evidence of James Kerr, medical officer LCC (education), ibid., vol. 1, 448.

p. 26 *Take their fingerprints* . . . ibid., 436.

p. 28 *Discover their family 'pedigree'* . . . Thomson, *Problem*, 246.

p. 28 *100,000 London schoolchildren too undernourished to benefit from schooling* . . . evidence of Dr Alfred Eichholz, Radnor Commission, *Minutes of Evidence*, vol. 1, 211.

p. 29 *In every case of alleged progressive hereditary* . . . *Report of the Inter-Departmental Committee on Physical Deterioration*, vol. 1, 14.

p. 29 *The plasticity of human material* . . . evidence of Alfred Eichholz, Inter-Departmental Committee on Physical Deterioration, *Minutes of Evidence*, vol. 2, 28.

p. 29 *The poorest and most ill-nurtured women* . . . ibid., 31.

Chapter 4

p. 31 *We are ceasing as a nation to breed intelligence* . . . Professor Karl Pearson, Huxley Lecture, 'The Laws of Inheritance in Man', 16 October 1903, published in *Science*, 13 November 1903.

p. 32 *The demographic decline of ruling castes* . . . Eversley, *Social Theories*, 38, quoted in Mazumdar, *Eugenics*, 24.

p. 32 *We cannot go on giving you health* . . . H.G. Wells, introduction to Sanger, *Pivot of Civilisation*.

p. 33 *Darwinists as no doubt opposing sunshine* . . . Arthur Newsholme, quoted in Gilbert, *Evolution*, 93.

p. 33 *We vainly looked for the keen competition between animals* . . . Kropotkin, *Mutual Aid*, 9.

p. 33 *Those communities which included the greatest number of the most sympathetic members* . . . Darwin, *Descent of Man*, 106–7.

p. 33 *The so-called fertility census* . . . Szreter, *Fertility*, 69–75.

p. 33 *Eugenicists had continued to undertake their own surveys* . . . *Eugenics Review*, July 1909, 87, quoted in Zedner, *Women*, 276.

p. 34 *Many of these women . . . are possessed of such erotic tendencies* . . . Tredgold, 'Feeble Minded', 720.

Chapter 5

p. 35 *It is an exception for any of them not to have been sexually tampered with* . . . evidence of Dr T.S. Clouston, Radnor Commission, *Minutes of Evidence*, vol. 3, 201.

p. 36 *What would be the almost inevitable result?* . . . evidence of Baldwyn Fleming, ibid., vol. 1, 115.

p. 36 *She is quite unable to protect herself* . . . ibid., vol. 1, 116.

p. 37 *She is on the borderline of imbecility* . . . evidence of Philip Bagenal, ibid., vol. 1, 130.

p. 37 *All [siblings] are of weak intellect* . . . evidence of Robert Parr, secretary of the NSPCC, ibid., vol. 2, 138.

p. 38 *Feeble-minded women are particularly open to the seductions of men* . . . evidence of Philip Bagenal, ibid., vol. 1, 132.

p. 38 *Again and again I have seen in the same district* . . . evidence of Dr J.S. Sutherland, ibid., vol. 3, 240.

p. 39 *Lost their character* . . . Maria Poole, secretary of the Metropolitan Association for Befriending Young Servants, ibid., vol. 1, 149.

p. 39 *I do not see myself what you could do* . . . ibid.

p. 40 *We are meeting almost every other day with cases* . . . evidence of Robert Parr, ibid., vol. 1, 142.

p. 40 *Trial of Frederick Briden, 18 November 1906* . . . Proceedings of the Old Bailey, 1674–1913 (online).

p. 41 *Some of them are girls who should never be allowed out of the house by themselves* . . . Ellen Pinsent, *The Lancet*, 21 February 1903.

p. 42 *It is the mild cases, which are capable of being well veneered* . . . Dendy, *Problem*, 19, quoted in Jackson, 'Institutional Provision', 67.

p. 43 *Motherhood gained a new dignity* . . . Hall, *Sex, Gender and Social Change*, 67–8.

p. 43 *Genetic purity and moral purity converged* . . . Bland, *Banishing the Beast*, 306.

p. 43 *New reality* . . . ibid., 229.

p. 44 *It is a byword among gynaecologists that 'Gynaecology is Gonorrhoea'* . . . Sayer letter to Home Secretary Reginald McKenna, 1 August 1912, National Archives, HO 45/10557/166505.

p. 45 *Never seen anything more interesting* . . . *Truth*, 7 October 1908.

p. 45 *As to real moral degenerates* . . . 'The Morally Defective Child: A Woman Doctor's Remarkable Lecture', *Daily Mail*, 30 September 1908.

p. 46 *Women concerned with the freedom of their sex are warned to keep a watchful eye* . . . 'Eugenics and Women', *The Vote*, 3 August 1912. I am grateful to Lucy Bland for this reference.

p. 47 *Women being shut up in asylums* . . . ibid., 7 November 1913.

p. 47 *With their ugly record of harsh and dictatorial dealings* . . . ibid., 15 June 1912.

p. 47 *Caird emphasised that the environment was a genuinely interactive force* . . . Richardson, *Love and Eugenics*, 203.

p. 47 *The chemical union of native bias with daily circumstance* . . . Mona Caird, 'Phases of Human Development – Morality', *Westminster Review*, January 1894. I am grateful to Angelique Richardson for this reference.

p. 48 *Why not include men?* . . . Hansard, House of Commons Debates, Fifth Series, vol. 56, 28 July 1913, col. 150.

p. 49 *She was so very tiresome* . . . evidence of Miss B. Walton Evans, Local Government Board inspector of boarded-out children, Radnor Commission, *Minutes of Evidence*, vol. 1, 150.

p. 49 *Exceptional gifts in certain directions and want of gifts in others* . . . ibid., vol. 1, 151.

p. 49 *Penitentiaries were intended as transformative institutions* . . . Mumm, '"Not Worse"', 527.

p. 50 *Not all sexually transgressive penitents had engaged in sexual activity voluntarily* . . . ibid., 529.

p. 50 *Every girl, however poor, lowborn, defective, unprotected, is precious in the sight of God* . . . *The Woman's Signal*, 28 June 1894, quoted in Hollis, *Ladies Elect*, 269.

Chapter 6

p. 55 *Boer War recruitment rejection rates* . . . *Report of the Inter-Departmental Committee on Physical Deterioration*, vol. 1, 7. Journalist Arnold White made the original allegation in the *Weekly Sun*, 28 July 1900. He stated that at the Manchester recruitment depot, of 12,000 young men who had wanted to enlist, 8,000 were rejected, and only 1,200 were found to be in peak fitness in all respects.

p. 55 *So far as the Committee are in a position to judge* . . . ibid., vol. 1, 46.

p. 56 *I think that the statement [by Pearson] is a pure assumption* . . . ibid., vol. 1, 39.

p. 56 *Disjointed and partial inquiries have taken place from time to time* . . . ibid., vol. 1, 3.

p. 56 *The aim of this Royal Commission was primarily to settle an administrative problem* . . . Owen, *Social Darwinism*, 92.

p. 57 *The adoption of heredity as the major influential factor in the causation of mental defect* . . . ibid., 154.

p. 59 *The measure should be classed as non-contentious* . . . 'Sterilisation of Eugenic Cranks', *Daily Herald*, 24 June 1912.

p. 60 *The old order, the old bland world* . . . Dangerfield, *Strange Death*, 67.

p. 61 *The present conspiracy of silence* . . . Sybil Gotto, 'Scheme of Organisation', *Eugenics Education Society Papers*, B1, quoted in Mazumdar, *Eugenics*, 29.

p. 61 *A model of social activism* . . . Mazumdar, *Eugenics*, 25.

p. 61 *I cannot say I am hopeful about the near future* . . . Reverend William Inge, 'Some Moral Aspects of Eugenics', *Eugenics Review*, April 1909, quoted in Mazumdar, *Eugenics*, 70.

p. 62 *The acknowledged dependants of the state* . . . Harris, *William Beveridge*, 119.

p. 62 *They incited guardians of the poor and educational staff to petition parliament* . . . Mazumdar, *Eugenics*, 16.

p. 62 *At least 120,000 or 130,000 feeble-minded persons at large in our midst* . . . *The Times*, 16 July 1910.

p. 63 *I am drawn to this subject in spite of many Parliamentary misgivings* . . . memo to Sir Charles Edward Troup, 27 May 1910, National Archives, HO 144/1098/197900.

p. 63 *Indiana State Reformatory sterilisation statistics and dates* . . . Gugliotta, 'Dr Sharp'.

p. 64 *A monument of ignorance and hopeless mental confusion* . . . memo from Dr Bryan Donkin to Troup, 31 May 1910, National Archives, HO 144/1098/197900.

p. 64 *It will be possible to empower, in general terms, to authorise medical operations* . . . memo from Troup to Churchill, 27 July 1910, National Archives, HO 144/1098/197900.

p. 64 *The unnatural and increasingly rapid growth of the feeble-minded* . . . Asquith papers, MS 12, folios 224–8, cited in Gilbert, 'Leading Churchill Myths' (online).

p. 65 *Winston is also a strong eugenist* . . . Blunt, *My Diaries*, quoted in Gilbert, ibid.

p. 65 *Tramps and wastrels . . . proper labour colonies* . . . letter from Churchill to George V, 10 February 1910, quoted in Addison, *Churchill*, 123.

p. 66 *The improvement of the British breed is my aim in life* . . . quoted in Gilbert, 'Leading Churchill Myths'.

p. 67 *The whole family life has been decimated* . . . James O'Grady, *Hansard, House of Commons Debates*, Fifth Series, vol. 21, 23 February 1911, col. 2074.

p. 67 *Profound regret that the occurrence of such a flagrant injustice* . . . *Common Cause*, 16 March 1911.

p. 67 *I challenge any person . . . to tell me what that woman was sent to prison for* . . . Chesterton, 'Are We All Mad?'

p. 67 *The object of this Bill is to regularise the lives* . . . Gershom Stewart, *Hansard, House of Commons Debates*, Fifth Series, vol. 38, 17 May 1912, col. 144.

p. 68 *In America I have seen several institutions* . . . ibid., col. 1457.

p. 68 *What we really wish to do is to free the sufferers* . . . ibid., col. 1445.

p. 68 *To a large class of the feeble minded, the greatest misery to them is the responsibility of liberty* . . . ibid., col. 1459.

p. 68 *This sort of women [sic] coming in year after* . . . ibid., col. 1447.

p. 69 *The procedure provided is as lenient as may be to the taxpayer* . . . ibid., col. 1519.

p. 69 *Will be put into these homes by forcible detention* . . . ibid., col. 1458.

Chapter 7

p. 70 *The obstructionist in chief . . . handsome young Liberal-Anarchist* . . . C[icely] V[eronica] Wedgwood, *Last of the Radicals*, 84.

p. 70 *A charm which the years developed into one of his surest weapons* . . . ibid., 80.

p. 71 *I was nearly off my head at the time* . . . Josiah Wedgwood, *Memoirs*, 84–5.

p. 72 *It is this government by specialists that you cannot argue with* . . . Josiah Wedgwood, *Hansard, House of Commons Debates*, Fifth Series, vol. 38, 17 May 1912, col. 1471.

p. 72 *You may depend upon this: if ever you have special doctors, they will shut up people by the score* . . . Lord Shaftesbury, *Minutes of Evidence of the Select Committee on Lunacy Law*, 545.

p. 73 *Men of ordinary intelligence* . . . 'Game of Obstruction', *Newcastle Daily Chronicle*, 29 May 1913.

p. 73 *At no time could you see more than forty members* . . . *Newcastle Daily Chronicle*, 29 May 1913.

p. 73 *The spirit at the back of the Bill is not the spirit of charity* . . . Josiah Wedgwood, *Hansard, House of Commons Debates*, Fifth Series, vol. 38, 17 May 1912, col. 1474.

p. 74 *It is not by building up temples of liberty* . . . Charles McCurdy, *Hansard, House of Commons Debates*, Fifth Series, vol. 53, 3 June 1913, col. 834.

p. 74 *The kind of liberty that at present is represented by being in and out of the [workhouse] casual wards* . . . Frederick Cawley, *Hansard, House of Commons Debates*, Fifth Series, vol. 41, 19 July 1912, col. 732.

p. 74 *It is a curious perversion of the desire for justice and liberty* . . . Mary Dendy, letter to *The Times*, 3 June 1912.

p. 75 *The NSPCC found that feeble-minded parents were less likely to mistreat their children than 'normal' parents* . . . Report of the Departmental Committee on Sterilisation ('The Brock Report'), 92.

p. 75 *We hear of liberty of the subject* . . . Will Crooks, Hansard, House of Commons Debates, Fifth Series, vol. 38, 17 May 1912, col. 1500.

p. 76 *The breeding of degenerate hordes of a demoralized 'residuum' unfit for social life* . . . Webb, *Difficulties*, 6.

p. 76 *Wedgwood and collectivism* . . . Mulvey, 'Individualist Thought', 29.

p. 77 *The relieving officer is already a terror in the poorer classes of society* . . . Josiah Wedgwood, Hansard, House of Commons Debates, Fifth Series, vol. 38, 17 May 1912, col. 1473.

p. 77 *I do think it shows an absolute want of understanding* . . . ibid., col. 1472.

p. 77 *A new and better society* . . . Wedgwood, *Last of the Radicals*, 95.

p. 77 *Combat the encroachment of the bureaucracy* . . . quoted in Jackson, *Borderland of Imbecility*, 226, n.14.

p. 78 *It remained active and united for nearly two months* . . . Wedgwood, *Last of the Radicals*, 94.

p. 78 *If these people are to be permitted to inflict their salacious rubbish on society* . . . *The Socialist*, 1 September 1913.

p. 78 *It gives the capitalist through his various legislative agencies* . . . ibid.

p. 78 *Labour Party leaders and the Mental Deficiency Act* . . . Jones, *Social Hygiene*, 58.

p. 78 *This conference expresses its approval* . . . *Daily Herald*, 31 January 1913.

p. 79 *Public opinion, which was everywhere (outside official circles) perceptible* . . . *Daily Herald*, 21 April 1913.

p. 79 *The outcome of poverty, low wages, bad housing* . . . *The Labour Leader*, 13 March 1913.

p. 80 *'Feeble-mindedness' is a new phrase under which you might segregate anybody* . . . Chesterton, *Eugenics and Other Evils*, 50.

p. 80 *Man has ceased to be a creature of God* . . . A.P. Mooney, *The Month*, September 1912, 276.

p. 80 *persons who from an early age display some permanent mental defect* . . . 3 & 4 Geo. 5., ch. 28, Mental Deficiency Act, 1–2.

p. 83 *This Bill is principally applied to women* . . . Hansard, House of Commons Debates, Fifth Series, vol. 38, 17 May 1912, col. 1472.

p. 83 *This Bill is eminently a Bill which we as men have no right to pass* . . . op. cit., col. 1476.

p. 83 *The considerable numbers of feeble-minded mothers who go into the workhouses* . . . Hansard, House of Commons Debates, Fifth Series, vol. 39, 10 June 1912, col. 633.

p. 83 *In many cases, these women were not capable of being mothers and had not the ordinary maternal instinct of an animal* . . . Dickinson's comment at the committee stage of the Bill, *The Times*, 13 November 1912.

p. 84 *Since individual control over moral behaviour was believed to be a direct reflection of temperamental* . . . Thomson, *Problem*, 244–5.

p. 84 *Absence of remorse always distinguishes the moral imbecile* . . . Alfred Tredgold, 'Moral Imbecility', *The Practitioner*, vol. 49 (1917), quoted in Watson, *Moral Imbecile*, 80.

p. 84 *A question which it is very difficult to decide* . . . Tredgold, *Mental Deficiency*, 295.

Chapter 8

p. 89 *Goodbye, lad. We hope you will soon settle in* . . . Barron, assisted by Banks, *Price to be Born*, 44.

p. 89 *No fault of theirs* . . . ibid., 45.

p. 91 *I saw many patients sent to Rampton* . . . ibid., 50.

p. 91 *The things he wanted me to do* . . . ibid., 76.

p. 91 *Having homosexuality forced on us* . . . ibid., 104.

p. 91 *These attentions made it more or less a full-time job* . . . ibid., 65.

p. 93 *Care should be taken* . . . Board of Control Committee on Mental Deficiency Colonies, Headley Committee report on the Arrangement of Colonies for Mental Defectives under the Mental Deficiency Acts, 1913–1927, National Archives, MH 58/97, 7.

p. 93 *The future of the race . . . a cross between public schools and guild training colleges* . . . 'Care of Mental Defectives', *The Times*, 19 March 1927.

p. 93 *Dangerous or objectionable habits* . . . Headley Committee report, op. cit.

p. 94 *Board of Control cost twenty-four times the sum spent on its predecessor body* . . . Treasury memo, 26 January 1922, National Archives, T 161/149.

p. 96 *Sandhill is not intended for young women of immoral tendencies* . . . letter dated 29 August 1922, quoted in Atkinson, Jackson and Walmsley, *Forgotten Lives*, 117.

p. 97 *We weren't able to mix with any of the female patients* . . . Potts and Fido, 'A Fit Person to be Removed', 110. A comprehensive illustrated history of Meanwood Park can be found at Mark M. Davis's website: http://www.meanwoodpark.co.uk/insight/hospital-tour-a-look-around-meanwood-park/

p. 97 *All the girls used to have to sit on one side* . . . ibid., 111.

p. 98 *The villas used to get very noisy* . . . ibid., 41.

p. 98 *Villa 8, it were a terrible villa* . . . ibid., 64.

p. 99 *Ensure clothing is not of unduly expensive quality* . . . ibid., 39.

p. 99 *Colony punishments* ... National Council for Civil Liberties, *50,000 Outside the Law*, 23.

p. 100 *Deceitful* ... Potts and Fido, *'A Fit Person to be Removed'*, 23.

p. 100 *The nurses beat me and hurt me for nothing* ... *Sunday Pictorial*, 23 April 1961.

p. 100 *I talked to an enchanting little boy of about five* ... *The Times*, 5 September 1949.

p. 101 *Peter Whitehead's experiences at Besford Court* ... Roxan, *Sentenced Without Cause*, 26–37.

p. 102 *I went to the grave of the teacher* ... 'A School of Hard Knocks', *Newcastle Evening Chronicle*, 11 March 2004, https://www.chroniclelive.co.uk/news/north-east-news/a-school-of-hard-knocks-1609856, accessed 6 February 2023.

p. 103 *The institutions procured magistrates to certify sufficient females as mentally defective* ... Platts-Mills, *Muck, Silk and Socialism*, 377.

p. 103 *An institution which takes all types and ages is economical* ... *Report of the Mental Deficiency Committee ('The Wood Report')*, 22–4.

p. 103 *The Act of 1913 gave arbitrary powers of detention* ... Platts-Mills, *Muck, Silk and Socialism*, 379.

p. 104 *'Tragedy of Tortured Girl: State Slaves Six Shillings a Year. Dora Thorpe Must Be Freed'* ... *John Bull*, 22 August 1928.

p. 104 *The cases of 'Ivy' and 'Jane'* ... *Civil Liberty: The Journal of the National Council for Civil Liberties*, 'Ivy', vol. 12, no. 4, spring 1955, 4–5, and 'Jane', vol. 10, no. 3, April 1950, 11.

p. 105 *St Catherine's farming output* ... memoranda prepared ahead of the 1940 informal visit to St Catherine's by the Board of Control, National Archives, MH 51/453.

p. 105 *Should be educational and not simple drudgery* ... letter from the National Union of County Officers dated 5 October 1942, 'Employment of Mental Defectives in their Colonies', National Archives, MH 51/457.

p. 105 *The cases of Henry C—, Violet B—, Cyril C— and the colony staffed by inmates* ... ibid.

p. 106 *In theory this should be full of pitfalls* ... ibid.

Chapter 9

p. 107 *All details from the Clifton Hampden case* ... National Archives, MH 85/167.

p. 108 *Girls little and big were admitted into friendship at once* ... *Strand*, April 1898.

p. 115 *Hundreds of them* ... Hill, *Starcross Hospital*, 22. David King's original study about moving to care in the community was published by the Nuffield Trust in 1991; King was head of the Exeter Health Authority throughout this period of institutional closures, writes Hill, and set in train a project to create an oral archive, undertaken between 1988 and 1991.

p. 116 *The cases of May R— and Catherine C—* ... papers of the London Association for the Care of the Mentally Defective, London Metropolitan Archives, LCC/PH/MENT/05/002.

p. 116 *Eliza H—, three illegitimate children* ... minutes of the London County Council Mental Deficiency Act Sub-Committee, May–July 1914, London Metropolitan Archives, LCC/MIN/720.

p. 117 *They would drift into the hands of the police* ... letter from the chair of the Somerset Quarter Sessions to Sir Robert S. Horne MP, Chancellor of the Exchequer, dated 4 January 1922, National Archives, T 161/149.

p. 118 *The female defectives sent to this institution consist almost entirely of young women who have given birth to illegitimate children* ... quoted in Atkinson, Jackson and Walmsley, *Forgotten Lives*, 117.

p. 118 *E.M., Liverpool, age 14. This girl was notified ... by the Liverpool Education Committee* ... letter to the Treasury from the West Lancashire Association for the Care of the Mentally Defective, National Archives, T 161/149.

p. 119 *In cases of girls of the prostitute class, it might well happen* ... undated memorandum, author Cyril Burt?, London Metropolitan Archives, LCC/PH/MENT/01/005.

p. 120 *The Bill had some merits* ... Wedgwood, *Memoirs*, 84.

p. 120 *Women and girls known or alleged to be immoral* ... London Metropolitan Archives LCC/PH/MENT/01/005.

p. 122 *The prevalence of sexual irregularities and illegitimacy on the part of the parents and relatives* ... Burt, 'Causes', 255.

p. 122 *Where all ages and both sexes are huddled together* ... ibid., 256.

p. 123 *The most general characteristic* ... ibid., 262.

p. 123 *Prostitutes' 'temperamental deficiency'* ... ibid., 265.

p. 123 *The feeble-minded prostitute is now a thing of the past* ... Mental Defectives and Crime, Reports for Years 1931 and 1932, National Archives, MH 51/454.

p. 123 *All these girls gave a history of a 'difficult childhood* ... Fairfield, 'Women Mental Defectives', 111.

p. 124 *Grow up to be dissolute women* ... quoted in Mahood, *Policing Gender*, 264–6.

p. 124 *Although in civilised communities regarded with almost universal condemnation* ... Flügel, *Psycho-Analytic Study*, 154–5. The source of Flügel's Chicago data: Chicago Vice Commission, *Social Evil in Chicago*.

p. 125 *The birth of the baby very often brings into prominence* ... memo from Stafford Cripps of the Society of Labour Lawyers to Minister of Health Aneurin Bevan, dated 10 October 1950, National Archives, LCO 2/3654. All remaining quotations in this chapter are from the same source.

Chapter 10

p. 127 *All details on John D—'s case* . . . National Archives, MH 86/11.

p. 131 *The place of Rampton in our mental deficiency system is causing grave concern* . . . Roxan, *Sentenced Without Cause*, 252.

p. 131 *At which both sexes are present under suitable supervision* . . . memoranda prepared ahead of the 1940 informal visit to Dovenby Hall by the Board of Control, National Archives, MH 51/453.

p. 133 *So-called kleptomaniacs . . . very often, they were brilliant people* . . . Radnor Commission, *Minutes of Evidence*, vol. 1, 361.

p. 133 *Just as tuberculosis is now known* . . . Dr Millais Culpin, 'The Present Position of Psychotherapy', *The Lancet*, 24 September 1921.

p. 134 *All details on Doreen S—'s case* . . . National Archives, MH 85/152.

p. 136 *All details on Jessie H—'s case* . . . National Archives, MH 85/143.

p. 136 *The history of Farmfield Colony* . . . National Archives, MH 51/453.

Chapter 11

p. 140 *the names of the same families kept cropping up* . . . Blacker, reviewing Lewis's work in 'Problem of Mental Deficiency', 369.

p. 140 *Eight per thousand mentally defective in England and Wales in the 1920s* . . . Report of the Mental Deficiency Committee ('The Wood Report'), 75.

p. 141 *'Social science intelligentsia'* . . . phrase coined by Starkey, 'Feckless Mother', 541.

p. 141 *Those families who, for their own wellbeing* . . . Tomlinson, *Families in Trouble*, 11.

p. 141 *The data are insufficient for definite conclusions* . . . ibid., 25.

p. 141 *There is some evidence that the persons included in the pedigrees* . . . Lidbetter, *Heredity*, 18.

p. 142 *shelved after this first volume* . . . Macnicol, 'In Pursuit of the Underclass', 308.

p. 142 *From their appearance they are strangers to soap and water* . . . Wofinden, 'Problem Families', 136; quoted in Tomlinson, *Families in Trouble*, 10.

p. 144 *As might be expected . . . a poor-type mother* . . . Tomlinson, *Families in Trouble*, 30.

p. 145 *A slum area has been cleared* . . . *The Wood Report*, 83.

p. 145 *The effect of the evacuation was to flood the dark places with light* . . . Hygiene Committee of the Women's Group on Public Welfare, *Our Towns*, xiii.

p. 146 *Public opinion is less tolerant of persistent lousiness* . . . Tomlinson, *Families in Trouble*, 1.

p. 146 *If the nation is seriously concerned to abolish poverty* . . . Seebohm Rowntree, preface to Pacifist Service Units, *Problem Families*.

p. 147 *The fetishisation of middle-class mores* . . . Macnicol, 'In Pursuit of the Underclass', 299.

p. 147 *It is not usually believed by those who study mental deficiency* . . . Penrose, *Mental Defect*, 57–8, quoted in Mazumdar, *Eugenics*, 162–3.

p. 148 *The rural mind has been taught caution in a hard school* . . . G.K. Chesterton, 'A Memorandum on the Report of the Committee on Mental Deficiency', part of a letter from the Distributive League to Labour MP and Minister of Health Arthur Greenwood, dated September 1929, National Archives, ED 50/124.

p. 149 *Migration has left behind a population inferior in mental quality* . . . *The Wood Report*, 81.

p. 150 *Arithmetic is not a test of IQ* . . . Chesterton memorandum, op. cit.

p. 150 *Stressing the continuous distribution of intelligence* . . . Jackson, *Borderland of Imbecility*, 176.

p. 150 *Decisions about which children were certified as mentally defective* . . . Sutherland, *Ability, Merit and Measurement*, 62.

p. 150 *Our examination of intelligence cannot take account of all those qualities* . . . Binet quoted in Board of Education *Report of the Consultative Committee on Psychological Tests*, 93.

p. 151 *They have a pseudo-scientific appearance of infallibility* . . . *ABPC News Letter* (Harpenden: Association of Parents of Backward Children, February 1950), vol. 2, no. 7, 5.

p. 151 *I myself have been present when tests have been given* . . . ibid. (January 1949), vol. 2, no. 1, 4.

p. 151 *Mrs C—'s son's failure to complete the test on time* . . . ibid., 7.

p. 151 *Our fears were dismissed as the unreasoning resistance of Doting Mammas* . . . ibid., February 1950, 5.

p. 152 *Suck it and see* . . . Arden, *Child of a System*, 31.

p. 152 *I have never known true happiness* . . . *Nottingham Evening Post*, 8 January 1949.

p. 153 *I read the sad story of James Smith* . . . Editor's Letter Bag, *Nottingham Evening Post*, 11 January 1949.

p. 155 *Burt's table of IQ and social standing* . . . Burt, 'Ability and Income', 83–4.

p. 155 *Lancelot Hogben and his Social Biology Group laboratory* . . . Kevles, *In the Name of Eugenics*, 130.

p. 155 *Characteristics that 'make for worldly success in business or professional life'* . . . Carl Murchison quoted in ibid., 144.

Chapter 12

p. 157 *From time to time you send a Visitor* . . . all details of this case, National Archives, MH 85/166.

p. 160 *A number of most villainous gangs* . . . Ives, *History*, quoted in Cook, *London*, 65.

p. 161 *purging sexual dissidence from public space* . . . Houlbrook, *Queer London*, 25.

p. 161 *I remember feeling sickened by what we did to him* . . . Tommy Dickinson, *Curing Queers*, 123.

p. 164 *In London, the number of males prosecuted for importuning other men for sex stood at around eighty each year* . . . Houlbrook, *Queer London*, 31.

p. 164 *1923 London magistrates court data* . . . London Metropolitan Archives, LCC/PH/MENT/01/007.

p. 164 *Crime statistics for the years 1931 to 1936 inclusive* . . . Board of Control surveys of mental deficiency and crime, National Archives, MH 51/454 and MH 51/456.

p. 165 *In 1931, of 589,657 people found guilty of criminal offences, 897 were declared mentally defective* . . . National Archives, MH 51/455.

p. 165 *Between the ages of sixteen and twenty-one, defectives were twice as likely to offend as 'normal' people* . . . ibid. The respective figures were 47.5 per cent against 21 per cent. Over the age of thirty, this ratio was reversed, with the 'normal' population representing 33 per cent of those found guilty, and alleged defectives at 13.5 per cent. Between the ages of twenty-one and thirty, offending rates were the same.

Chapter 13

p. 167 *5,240 certified mental defectives are free* . . . *Daily Herald*, 29 October 1951.

p. 167 *He was 'a real problem'* . . . Fairfield, *Trial*, 2.

p. 168 *No 'harmful characteristics'* . . . ibid., 3.

p. 168 *A tousle-haired boy with thin features* . . . Roxan, *Sentenced Without Cause*, 28.

p. 168 *What would you do if I killed you?* . . . Fairfield, *Trial*, 124.

p. 168 *Not of violent or dangerous propensities* . . . ibid., 6.

p. 170 *Mental defectives are by definition NOT insane* . . . ibid., 3.

p. 170 *No decisive test which would place the high-grade defective once and for all outside the category of normal people* . . . ibid., 23.

p. 171 *No one has yet been clever enough* . . . ibid., 28.

p. 171 *Vindictive violence directed against some creature who cannot retaliate* . . . ibid., 29.

p. 172 *He derived from it some emotional gratification* . . . Fairfield, *Trial*, 28.

p. 172 *Something came over me in my head* . . . details of the Lenchitsky case, National Archives, CRIM 1/2361.

p. 174 *Details of Mrs J—'s sons* . . . National Archives, MH 88/29.

p. 175 *Magistrates 'often forced into decisions which they themselves deplore'* . . . Irene Ward, letter to Rab Butler, 24 October 1959, ibid.

p. 175 *I do not see why a young girl's life should be worried* . . . Irene Ward, letter to Derek Walker-Smith at the Ministry of Health, 29 October 1959, ibid.

p. 176 *The gravity of a sexual offence is masked* . . . Report of the Departmental Committee on Sexual Offences Against Young Persons, 9.

p. 176 *In many children, the commission of the offence causes a shock* . . . ibid., 66.

p. 177 *To awaken or excite the dormant passions* . . . ibid.

p. 177 *The superstition that connection with a virgin will cure a man of venereal disease* . . . ibid., 13.

p. 178 *We have had such varied types of offenders brought to our notice* . . . ibid., 54.

p. 178 *Other expert evidence confirms these statistics* . . . ibid., 56.

p. 178 *Very many years, and no punishment appears to have acted as a deterrent* . . . ibid., 60.

p. 178 *Prolonged detention of men who appear quite incapable of abstaining from indecent exposure* . . . ibid., 60.

p. 178 *It sets a wrong moral tone* . . . ibid., 63.

p. 179 *High-grade feeble-minded persons are exceedingly difficult to detect* . . . ibid., 57.

Chapter 14

p. 183 *He is a typical example of the village fool with dangerous sexual tendencies* . . . letter from John Barker at the Home Office, 7 January 1911, National Archives, HO 144/1088.

p. 183 *A certain learned judge* . . . National Archives, MH 51/454.

p. 184 *Sterilisation would help greatly towards reducing the grave state of mental defectiveness* . . . The Times, 13 July 1929.

p. 184 *Details on the castrations at the High Teams Institution, Gateshead* . . . National Archives, MH 51/712 and MH 79/291.

p. 185 *She has now begun to menstruate* . . . National Archives, MH 51/172.

p. 185 *He worked very well for me when he was at home* . . . ibid.

p. 186 *Difficult 'to say with certainty that the genetic endowment of any individual'* . . . The Brock Report, 36.

p. 186 *Mentally defective and mentally disordered parents are, as a class, unable to discharge their social and economic liabilities* . . . ibid., 38.

p. 186 *Most 'low-grade' cases could never live independently* . . . ibid., 29.

p. 186 *The unstable and antisocial defective remains unstable and antisocial* . . . ibid., 37.

p. 187 *Driving defect underground* . . . ibid.

p. 187 *So long as there is no unfair pressure* . . . ibid., 43.
p. 187 *If the doctor believed that the feeble-minded patient really hadn't understood the issue* . . . ibid., 47.
p. 187 *The poor man, a victim of an inherited physical or mental disorder* . . . ibid., 43.
p. 188 *So where had the left's appreciation of the allegedly 'progressive' nature of eugenics gone?* . . . I have drawn in these paragraphs upon the work of Jones, *Social Hygiene*; and Macnicol, 'Eugenics'.
p. 188 *It is from this section that all the striking and sensational cases are drawn* . . . Scientific Correspondent (said to be Hogben), 'Sterilisation of Defectives'.
p. 189 *Ancestor-worship, anti-semitism, colour prejudice, anti-feminism* . . . Hogben, *Genetic Principles*, 209, cited in Macnicol, 'Eugenics', 153.
p. 189 *At this point in the early 1930s, with the tide of left-wing criticism setting against it* . . . Mazumdar, *Eugenics*, 155.
p. 189 *We find a remarkable consensus of opinion* . . . The Brock Report, 13.
p. 189 *It is impossible in the present state of our knowledge* . . . ibid., 21.
p. 189 *We find ourselves compelled to the conclusion* . . . ibid., 27.
p. 190 *I am afraid that I tend to be rather prejudiced on this subject* . . . Eustace Percy, letter to Riddell, 6 May 1929, National Archives, ED 50/124.
p. 190 *We are getting a larger and larger quantity of human dregs* . . . Professor Edward East, quoted in Riddell's lecture text, 'The Sterilization of the Unfit', in Riddell's reply to Percy, 8 May 1929, National Archives, ED 50/124.
p. 190 *The truth is that we are spending far too much on these unfortunate people* . . . Riddell's letter, 8 May 1929, op. cit.
p. 191 *It looks as if we are going to be eaten out of house and home by lunatics* . . . Riddell's letter to Neville Chamberlain, 27 April 1929, National Archives, MH 58/103, quoted in Macnicol, 'Eugenics', 151.
p. 191 *When we have all seen for ourselves, as I have* . . . Dr Richard Berry, 'The National Lethal Chamber', letter to *The Times*, 6 February 1930.
p. 191 *I do not share your views as to the 'sanctity of human life'* . . . Dr Richard Berry, letter to the *Eugenics Review*, April 1930.
p. 191 *Extreme proposals always do worse than defeat their own ends* . . . editorial in ibid.
p. 193 *Give an official the right to forbid, and he begins to think he has the right to command* . . . Gerald Gould, 'Not Fit to Have Babies: Perils of Having Dictators to Control Our Lives', *Sunday Chronicle*, 23 June 1929.
p. 193 *Mental deficiency as a problem affecting the state is of purely modern origin* . . . Robins, 'Mental Deficiency'.
p. 194 *'Self-made men' are especially dangerous to the community* . . . Hamilton Fyfe, *Reynolds's Illustrated News*, 11 February 1934.

p. 194 *Segregation has not yet been seriously tried* . . . Reynolds's Illustrated News, 14 January 1934.

p. 194 *Sterilisation of the unfit is an initial step towards selective breeding* . . . Gibson quoted in Macnicol, 'Eugenics', 163.

p. 195 *Blind faith in doctors remained a problem and 'must be broken down* . . . minutes of the National Council for Lunacy Reform 1931 AGM, Wellcome Collection Archives, SA/MIN/A/1/3.

p. 195 *That it is essential to curtail the power of the Board of Control* . . . minutes of the National Council for Lunacy Reform 1933 AGM, Wellcome Collection Archives, SA/MIN/A/1/3.

p. 195 *Before the General Election of 1935* . . . Macnicol, 'Eugenics', 158.

p. 196 *We have no right to take liberties to experiment upon people* . . . quoted in Fennell, Treatment Without Consent, 258.

p. 196 *Labour MPs who in the early 1930s were sympathetic to voluntary sterilisation* . . . Jones, Social Hygiene, 58.

p. 196 *Common to see statements linking such a procedure to a lessening of male sexual violence* . . . Macnicol, 'Eugenics', 164.

p. 196 *This conference, believing it is wrong to condemn people to a choice between unnatural abstinence* . . . Report of the Seventeenth National Conference of Labour Women, 92.

Chapter 15

p. 198 *On the edge of the beautiful Valley of the Moon* . . . Hodson, Human Sterilization Today, 22.

p. 199 *The total number of people in American institutions who had been sterilised* . . . The Brock Report, 35.

p. 199 *Patients who subsequently married have shown considerable resentment* . . . National Association for Mental Health (NAMH) Mental Deficiency Sub-Committee, 'Notes on Sterilisation, 1951', Wellcome Collection Archives, SA/MIN/A/2/6.

p. 199 *In our view, there is no justification for sterilising defectives who are unfit for community life* . . . The Brock Report, 36.

p. 200 *A most charming and sympathetic woman* . . . Hodson, Human Sterilization Today, 20.

p. 200 *Some thirty US states had passed laws banning certain types of individuals from marrying, on eugenic grounds* . . . Kevles, In the Name of Eugenics, 100.

p. 201 *Belong to the shiftless, ignorant and worthless class of anti-social whites of the South* . . . details of the Buck case, ibid., 110–11.

p. 201 *I never knew anything about it* . . . 'I'm not mad, just broken-hearted', interview by Gary Robertson, *Richmond Times-Dispatch*, 23 February 1980, quoted in Noll, *Feeble-Minded in Our Midst*, 73.

p. 202 *It is better for all the world* . . . Noll, *Feeble-Minded in Our Midst*, 71.

p. 202 *Three thousand people from the various undesirable categories were being sterilised each year* . . . ibid., 72.

p. 203 *Easy prey to the sexual aggressions of males of superior intellect* . . . quoted in ibid., 75.

p. 203 *Almost half of the sterilised were black* . . . Kevles, *In the Name of Eugenics*, 168.

p. 203 *In 1923, 100,000 women had petitioned, via the Women's National Council* . . . Bent Sigurd Hansen, 'Something Rotten in the State of Denmark: Eugenics and the Ascent of the Welfare State', in Broberg and Roll-Hansen (eds.), *Eugenics*, 31.

p. 204 *The last woman left the island in 1961* . . . Graham, 'Rigid and Hard Lives'; Birgit Kirkebaek, 'Sexuality as Disability'.

p. 204 *The second largest group were 'women exhausted from many childbirths'* . . . National Association for Mental Health (NAMH) Mental Deficiency Sub-Committee, 'Notes on Sterilisation, 1951', Wellcome Collection Archives, SA/MIN/A/2/6.

p. 205 *An anti-social way of life* . . . Gunnar Broberg and Mattias Tyden, 'Eugenics in Sweden: Efficient Care', in Broberg and Roll-Hansen (eds.), *Eugenics*, 108.

p. 205 *One report put their number at between eight and ten per cent of all Germans* . . . Adams (ed.), *Well-Born Science*, 38.

p. 206 *Many far-sighted men and women in both England and America have been working earnestly* . . . Noll, *Feeble-Minded in Our Midst*, 72.

p. 206 *Between 200,000 and 400,000 Germans were compulsorily sterilised* . . . Adams (ed.), *Well-Born Science*, 44.

p. 206 *An estimated 7,000–10,000 children and young people killed, usually by lethal injection* . . . Burleigh, 'Psychiatry', 222; The Holocaust Encyclopedia, 'Euthanasia Program and Aktion T4' (online).

p. 207 *Odd assortment of highly educated, morally dulled humanity* . . . Burleigh, 'Psychiatry', 223–4.

p. 207 *The vast majority of members of the Eugenics Society and its associated commissions were revolted by the racism of Nazi eugenics* . . . Mazumdar, 'Eugenics', 37.

p. 208 *Tarred with the same brush* . . . Lancelot Hogben, quoted in Macnicol, 'Eugenics', 151.

p. 208 *Doubly deplorable . . . race warfare* . . . *Eugenics Review*, July 1933.

p. 208 *From pulpit and platform* . . . ibid., October 1933.

p. 208 *Details on the International Federation of Eugenic Organizations* . . . Hart, 'Watching'.

p. 209 *I feel inclined to suggest minuting this 'Bring up again on Jan 1st 1950'* . . . memo to Churchill from Charles Masterman, 20 September 1910, National Archives, HO 144/1098/197900.

Chapter 16

p. 210 *Hiding like a hare from the hounds* . . . Evening Standard, 18 January 1935.

p. 210 *Practically all the girls had sexual mania* . . . Harman, *Sylvia Townsend Warner*, 99–100.

p. 211 *It is our considered opinion that neither Miss Stevenson nor Mrs Stevenson are suitable persons* . . . details of the Chaldon Vicarage affair, National Archives, MH 58/74.

p. 213 *Sometimes the daughters of tramps* . . . report of the libel trial, *Evening Standard*, 18 January 1935.

p. 214 *Bad-tempered, violent, vicious, irritable, quarrelsome* . . . 'Disease and Intelligence', in British Psychological Society, 'Report of Symposium', 27; Ruiz, 'Disease'.

p. 214 *Compulsory segregation put the erratic at the mercy of the fanatic* . . . *The Times*, 19 March 1927.

p. 215 *In 1929, over 44,000 people were living under guardianship, statutory supervision or voluntary supervision* . . . *Report of the Royal Commission on the Laws Relating to Mental Illness and Mental Deficiency*, appendix D, 'Number of mental defectives under care, 1929–1952'.

p. 215 *In 1932, the Brighton Guardianship Society was criticised by the local council* . . . Westwood, 'Care in the Community', 67.

p. 216 *The importation of steadily increasing numbers of defectives into Brighton* . . . ibid., 68.

p. 216 *The present restrictions are handicapping my social life in every way* . . . Brewster's letter to the Board of Control, 23 February 1931, National Archives, MH 85/147.

p. 217 *I consider it a great injustice that you still keep him tied by petty restrictions* . . . J.S. Copsey's letter, dated 10 February 1932, ibid.

p. 218 *After lingering for years in the ghastly asylum* . . . *John Bull*, 16 July 1932.

p. 218 *It was while I was there that I met a young soldier* . . . 'Marjorie at 33 Is Free for the First Time', *News of the World*, 22 October 1950.

p. 219 *Lost touch with her baby's father* . . . National Council for Civil Liberties, *50,000 Outside the Law*, 6.

p. 219 *Mildred M—'s case* . . . ibid., 26.

p. 220 *You are not allowed to speak to MEN* . . . Rolph, *History*, 177. Rolph's is the first work to examine in depth the role of the hostel in England with regard to mental deficiency.

p. 220 *The women's lives were monitored* . . . ibid., 176.

p. 220 *The depth of some of the friendships* ... ibid., 229.
p. 220 *Blofield Hall hostel for males, friendships with work-mates* ... ibid., 246.
p. 220 *Did not know how to relate to the [Blofield] men* ... ibid., 299.
p. 221 *She knows nothing about work and the outside world* ... National Council for Civil Liberties archive, Hull History Centre, U DCL/83/6.
p. 221 *Homosexual practices ... associating with the opposite sex ... seen talking to a schoolgirl* ... Jan Walmsley, Dorothy Atkinson and Sheena Rolph, 'Community Care and Mental Deficiency, 1913 to 1945', in Bartlett and Wright (eds.), *Outside the Walls*, 195–6.
p. 221 *The Henry Cottingham case* ... *Sussex Express*, 15 July 1927.
p. 222 *The Violet A— case* ... *Hastings & St Leonard's Observer*, 29 April 1939 and 15 July 1939.
p. 222 *The George Bostock case* ... *Lancaster Guardian & Observer*, 20 October 1950.
p. 223 *This dangerous creature [having been] put into the world without reasonable care* ... *The Courier and Advertiser*, 10 June 1937.
p. 224 *The possibilities of exploitation are obvious* ... memorandum from the Society of Labour Lawyers to the Minister of Health, 10 October 1950, National Archives, LCO 2/3654.
p. 225 *Between 1951 and 1955 there had been 698 fatalities* ... *West Sussex Gazette*, 29 November 1956.
p. 225 *likely to restore many of these folk to full mental health* ... Mr Arnold, minute of Brighton Guardianship Society meeting on 24 October 1957, National Archives, MAF 258/27.
p. 225 *It seemed that the type of farmer* ... ibid.
p. 225 *She is to be brought out of freedom and into captivity* ... the case of Alma S—, National Archives, MH 51/457.
p. 226 *Operates as a sieve which sifts the wrong way* ... National Council for Civil Liberties, *50,000 Outside the Law*, 27.

Chapter 17

p. 229 *Courteous appreciation of points raised and a reasoned statement of reply* ... National Council for Civil Liberties, *50,000 Outside the Law*, 29.
p. 230 *It is an old stunt* ... letter dated 23 September 1951. Details of 'Jane's' case, National Council for Civil Liberties archive, Hull History Centre, U DCL/83/9.
p. 233 *But I think it was best, for somehow you appear to be a family that does not get on well together* ... Details of 'Glenda's' case, NCCL archive, U DCL/83/4.
p. 235 *Lazy ... obstinate ... extremely nervous and frightened* ... details of Jean P—'s case, NCCL archive, U DCL/83/8.
p. 236 *This was a perfect example of petty dictation in the lives of working-class people* ... letter dated 14 October 1953, in ibid.

p. 236 *They are so big and hard to fight* . . . selection of quotations from the letters in the NCCL archive.

p. 237 *Quite capable of adapting themselves to a normal life* . . . Elizabeth Allen's letter, *The Times*, 16 April 1952.

p. 237 *Insufficient educational facilities are being provided for this type of child* . . . *The Lancet*, 12 May 1951.

p. 238 *Many and devious are the means employed to obtain consent* . . . National Council for Civil Liberties, *50,000 Outside the Law*, 14.

p. 239 *Details of 'Bert's' case* . . . ibid., 16.

p. 239 *Details of 'John's' case* . . . *Civil Liberty*, vol. 12, no. 7, spring 1958, 11–12.

p. 239 *Details of the Kathleen Rutty case* . . . *The Times*, 7 February 1956.

p. 240 *No persons of any age were to be confined in institutions* . . . *The Times*, 10 March 1956.

p. 241 *under the pretence that they had been found neglected* . . . *The Western Mail*, 2 July 1956.

p. 241 *The false statements made by the authorities were routine and conventional in such cases* . . . Platts-Mills, *Muck, Silk and Socialism*, 378.

Chapter 18

p. 243 *Pathological defects or abnormalities of personality* . . . The Percy Commission, 52.

p. 243 *Psychopathic disorder* . . . *a persistent disorder or disability of the mind* . . . 7 & 8 Eliz. 2., ch. 72, Mental Health Act, section 4, clause 5.

p. 244 *The exception would be the promiscuous feeble-minded woman* . . . memo to the Ministry of Health, National Archives, MH 121/44.

p. 244 *Some thirty per cent of people certified as mental defectives had been in detention between ten and twenty years* . . . statistics on mental deficiency, as of 31 December 1954, National Archives, MH 121/44.

p. 245 *All of a sudden it seemed as though we got the directive* . . . Rolph, *History*, 133.

p. 245 *The date 1959 was emblazoned on my memory* . . . ibid., 288.

p. 245 *It were some big man up from London that sent for me* . . . Potts and Fido, 'A Fit Person to be Removed', 126–7.

p. 246 *The process would have been 'very frightening* . . . Atkinson, Jackson and Walmsley, *Forgotten Lives*, 34. Mabel later wrote her own autobiographical chapter, in Atkinson et al. (eds.), *Good Times, Bad Times: Women with Learning Difficulties Telling their Stories* (Kidderminster: BILD, 2000).

p. 246 *Right, you all have to go back in the hospital!* . . . Atkinson, Jackson and Walmsley, *Forgotten Lives*, 29.

p. 246 *We could be taken back to Whixley if we did any little thing wrong* . . . Barron, assisted by Banks, *A Price to Be Born*, 110.

p. 246 *I've got to watch I don't slip back into institutional ways* . . . ibid., 122.

p. 246 *Had her life absolutely ruined* . . . Jean Waldron, nursing sister, interview in Hill, *Starcross Hospital*, 22.

p. 247 *The popular press and general public often have a picture of patients* . . . Raphael and Peers, *Psychiatric Hospitals*, 14.

p. 247 *A hospital that can offer playing fields, dances, television, outings and holidays* . . . letter to the Ministry of Health, National Archives, MH 137/405.

p. 247 *We didn't want them to return home* . . . Rolph, *History*, 293.

p. 247 *The outside world was somewhere they wanted to get to* . . . ibid., 292.

p. 248 *Twenty per cent of the females and fifteen per cent of the males were not mental patients proper* . . . S. Smith, medical superintendent, 'Lancaster Moor, the Metamorphosis of a Mental Hospital', *The Lancet*, 10 September 1960. The superintendent added that there was a waiting list of over one thousand patients for mental deficiency beds, and Lancaster Moor had accepted fifty of these, even though they were out of the area. These patients were what he described as 'cot-and-chair' cases, in need of complete nursing, since they were physically as well as intellectually profoundly disabled.

p. 249 *Calderstones Hospital dilemma regarding two male sex offenders* . . . *The Lancet*, 2 February 1963.

p. 250 *She often jumped on other girls' bicycles and pedalled off laughing* . . . *Sunday Pictorial*, 16 April 1961.

p. 252 *All I want now is a little job doing domestic work* . . . *Daily Mirror*, 22 May 1972.

p. 252 *There were likely to be 30,000 people still being detained for 'social reasons' in mental hospitals* . . . *Birmingham Post*, 22 May 1972.

p. 252 *'Janet', thirteen years old, was discovered on an adult ward of Stanley Royd Mental Hospital in Wakefield* . . . *Daily Mirror*, 6 August 1972.

p. 253 *The nation was moving towards degeneration* . . . the full speech is to be found on the Margaret Thatcher Foundation website: Joseph, 'Speech at Edgbaston' (online).

p. 253 *Frighteningly irresponsible* . . . *Leicester Mercury*, 25 October 1974.

p. 253 *Denunciations of Joseph's speech* . . . letters page, *The Times*, 25 October 1974.

p. 254 *Unborn Babies Targeted in Crackdown on Criminality* . . . *The Guardian*, 16 May 2007.

p. 254 *Fewer subsequent pregnancies and greater intervals between births* . . . Department of Health, 'Family Nurse Partnership Programme Information Leaflet' (online), 7.

p. 254 *No evidence of benefit from FNP* . . . *The Lancet*, 13 October 2015.

Conclusion

p. 256 *100,000 or so girls and infants who passed through the Irish Magdalene Laundries* ... Final Report of the Commission of Investigation into Mother and Baby Homes (online).

p. 257 *The general public are too apathetic about these things* ... George Scott-Rimington, letter to the NCCL dated 7 May 1952, National Council for Civil Liberties archive, U DCL/83/9.

Afterword

p. 259 *The average length of stay is five and half years* ... Mencap, 'No Freedom' (online).

p. 260 *Over five and a half thousand incidents of mechanical or chemical restraint or seclusion were reported* ... NHS, 'Learning Disability Services Monthly Statistics' (online).

p. 260 *The Steven Neary case* ... Neary, 'Love, Belief, and Balls' (online); Essex Chambers, 'London Borough of Hillingdon v Neary' (online); Legal Action Group, 'The Court of Protection: Steven Neary's Story' (online).

p. 261 *The Adam Downs case* ... ITV News, 'Barbaric' (online).

p. 262 *Deprivation of Liberty Safeguards definition* ... Alzheimer's Society, 'Deprivation of Liberty Safeguards (DoLS)' (online).

p. 262 *The person is under continuous supervision and control and is not free to leave* ... ibid.

p. 263 *I have no doubt that this area of the law is so complex* ... OBITER J, 'Law and Lawyers: Deprivation of Liberty' (online).

p. 263 *The deprivation of liberty safeguards are so complicated* ... email communication with Lucy Series and Series' website, The Small Places. Her chapter, 'Of Powers and Safeguards', will appear in the forthcoming book *Re-Imagining Health and Care Law*, ed. Atina Krejewska and Jean McHale (Edward Elgar Publishing, 2024).

p. 263 *What good is it making someone safer if it merely makes them miserable?* ... Robson and Kitzinger, 'Inspired by Bournewood' (online).

p. 264 *Place the person at the heart of decision-making* ... Social Care Institute for Excellence, 'What Are Liberty Protection Safeguards?' (online).

p. 264 *You can't just exist in this extraordinary world* ... ibid.

Appendix 3

p. 272 *NAHM replies* ... 1951 minutes of NAMH Mental Deficiency Sub-Committee, Wellcome Library Archives, SA/MIN/A/2/6.[2]

Bibliography

Books, articles, theses and pamphlets

Adams, Mark B. (ed.), *The Well-Born Science: Eugenics in Germany, France, Brazil, and Russia* (New York: Oxford University Press, 1990)

Addison, Paul, *Churchill on the Home Front, 1900–1955* (London: Jonathan Cape, 1992)

Arden, Noele, *Child of a System* (London: Quartet, 1977)

Atkinson, Dorothy, Jackson, Mark and Walmsley, Jan, *Forgotten Lives, Exploring the History of Learning Disability* (Kidderminster: BILD Publications, 1997)

Barron, David, assisted by Edwin Banks, *A Price to be Born: My Childhood and Life in a Mental Institution* (Harrogate: Mencap, 1996)

Bartlett, Peter and Wright, David (eds.), *Outside the Walls of the Asylum: The History of Care in the Community, 1750–2000* (London: Athlone Press, 1999)

Blacker, C.P., 'The Problem of Mental Deficiency', *Political Quarterly*, vol. 1 (1930), no. 3, 368–85

Bland, Lucy, *Banishing the Beast: Feminism, Sex and Morality* (London: Tauris Parke, 2001)

Blunt, W.S., *My Diaries: 1888–1914* (New York: Alfred A. Knopf, 1921)

Bosanquet, Bernard (ed.), *Aspects of the Social Problem* (London: Macmillan, 1895)

British Psychological Society, 'Report of Symposium on the Definition and Diagnosis of Moral Imbecility', *British Journal of Medical Psychology*, vol. 6 (1926), no. 3, 27

Broberg, Gunnar and Roll-Hansen, Nils (eds.), *Eugenics and the Welfare State: Sterilisation Policy in Denmark, Sweden, Norway and Finland* (East Lansing: Michigan State University Press, 2005)

Burleigh, Michael, 'Psychiatry, German Society, and the Nazi "Euthanasia" Programme', *Social History of Medicine*, vol. 7 (1994), no. 2, 213–28

Burrows, George Mann, *Commentaries on the Causes, Forms, Symptoms and Treatment, Moral and Medical, of Insanity* (London: T. & G. Underwood, 1828)

Burt, Cyril, 'Ability and Income', *British Journal of Educational Psychology*, vol. 13 (1943), no. 2, 83–96

Burt, Cyril, 'The Causes of Sex Delinquency in Girls', *Health and Empire, the Quarterly Journal of the British Social Hygiene Council*, vol. 1 (1926), 251–71

Chesterton, G.K., 'Are We All Mad?', *The Daily News*, 18 February 1911

Chesterton, G.K., *Eugenics and Other Evils* (London: Cassell, 1922; new edition, Seattle: Inkling Books, 2000)

Chicago Vice Commission, *The Social Evil in Chicago* (Chicago: Gunthorp-Warren, 1911)

Cook, Matt, *London and the Culture of Homosexuality, 1885–1914* (Cambridge: Cambridge University Press, 2003)

Dangerfield, George, *The Strange Death of Liberal England* (London: MacGibbon & Kee, 1935)

Darwin, Charles, *The Descent of Man: And Selection in Relation to Sex* (London: John Murray, 1871)

Dendy, Mary, *The Problem of the Feeble Minded* (Manchester: Thomas Wyatt, 1911)

Dickinson, Tommy, *Curing Queers: Mental Nurses and Their Patients, 1935–1974* (Manchester: Manchester University Press, 2015)

Ellis, Havelock, *The Criminal* (London: Walter Scott, 1890)

Eversley, D.E.C., *Social Theories of Fertility and the Malthusian Debate* (Oxford: Clarendon, 1959)

Fairfield, Laetitia, *The Trial of John Thomas Straffen and Eric Fullbrook*, part of the Notable British Trials series (London: Hodge, 1954)

Fairfield, Laetitia, 'Women Mental Defectives and Crime', *The Lancet*, 10 January 1931

Fennell, Phil, *Treatment Without Consent: Law, Psychiatry and the Treatment of Mentally Disordered People since 1845* (London: Routledge, 1996)

Flügel, John Carl, *The Psycho-Analytic Study of the Family* (London: International Psycho-Analytical Press, 1921)

Galton, Francis, 'Eugenics, Its Definition, Scope and Aims', in *Essays in Eugenics* (London: Eugenics Education Society, 1909)

Galton, Francis, *Inquiries into Human Faculty and Its Development* (London: Macmillan, 1883)

Gilbert, Bentley, *The Evolution of National Insurance in Great Britain* (London: Michael Joseph, 1966, reissued 2019)

Graham, Jane, 'The Rigid and Hard Lives of the "Loose and Easy" Women on the Danish island of Sprogø', *The Copenhagen Post*, 4 December 2016

Gugliotta, Angela, 'Dr Sharp and His Little Knife: Therapeutic and Punitive Origins of Eugenic Vasectomy, Indiana 1892–1921', *Journal of the History of Medicine and Allied Sciences*, vol. 53 (1998), no. 4, 371–406

Hall, Lesley, *Sex, Gender and Social Change in Britain Since 1880* (New York: St Martin's Press, 1999)

Harman, Claire, *Sylvia Townsend Warner, A Biography* (London: Chatto & Windus, 1989)

Harris, José, *William Beveridge, A Biography* (Oxford: Clarendon, 1977)

Hart, Bradley W., 'Watching The "Eugenic Experiment" Unfold: The Mixed Views of British Eugenicists Toward Nazi Germany in the Early 1930s', *Journal of the History of Biology*, vol. 45 (2012), no. 1, 43–63

Hill, Caroline, *Starcross Hospital: What the Voices Tell Us* (Author, 2020)

Hodson, Cora B.S., *Human Sterilization Today, A Survey of the Present Position* (London: Watts, 1934)

Hogben, Lancelot, *Genetic Principles in Medicine and Social Science* (London: Williams & Norgate, 1931)

Hollis, Patricia, *Ladies Elect: Women in English Local Government, 1865–1914* (Oxford: Clarendon, 1987)

Houlbrook, Matt, *Queer London: Perils and Pleasures in the Sexual Metropolis, 1918–1957* (Chicago: University of Chicago Press, 2005)

Ives, George, *A History of Penal Methods: Criminals, Witches, Lunatics* (London: Stanley Paul & Co., 1914)

Jackson, Mark, *The Borderland of Imbecility: Medicine, Society and the Fabrication of the Feeble Mind in Late Victorian and Edwardian England* (Manchester: Manchester University Press, 2000)

Jackson, Mark, 'Institutional Provision for the Feeble Minded in Edwardian England: Sandlebridge and the Scientific Morality of Permanent Care', in Wright, David and Digby, Anne (eds.), *From Idiocy to Mental Deficiency: Historical Perspectives on People with Learning Disabilities* (London: Routledge, 1996), 271–94

Jones, Greta, *Social Hygiene in Twentieth-Century Britain* (London: Croom Helm, 1986)

Kevles, Daniel, *In the Name of Eugenics: Genetics and the Uses of Human Heredity* (Berkeley: University of California Press, 1985)

Kirkebaek, Birgit, 'Sexuality as Disability: The Women on Sprogø and Danish Society', *Scandinavian Journal of Disability Research*, vol. 7 (2005), no. 3–4, 194–205

Kropotkin, Peter, *Mutual Aid: A Factor of Evolution* (London: Heinemann, 1902)

Lidbetter, E.J., *Heredity and the Social Problem Group* (London: E. Arnold, 1933)

Macnicol, John, 'Eugenics and the Campaign for Voluntary Sterilization in Britain Between the Wars', *Social History of Medicine*, vol. 2 (1989), no. 2, 147–69

Macnicol, John, 'In Pursuit of the Underclass', *Journal of Social Policy*, vol. 16 (1987), no. 3, 293–318

Mahood, Linda, *Policing Gender, Class and Family: Britain 1850–1940* (London: Routledge, 1995)

Maudsley, Henry, *Responsibility in Mental Disease* (London: King & Co., 1874)

Mazumdar, Pauline, *Eugenics, Human Genetics and Human Failings: The Eugenics Society, Its Sources and Its Critics in Britain* (London: Routledge, 1992)

Mulvey, Paul, 'Individualist Thought and Radicalism: Josiah C. Wedgwood's Battle against the Collectivists, 1906–1914', *Journal of Liberal History*, vol. 56 (autumn 2007), 26–33

Mumm, Susan, '"Not Worse than Other Girls": The Convent-Based Rehabilitation of Fallen Women in Victorian Britain', *Journal of Social History*, vol. 29 (1996), no. 3, 527–47

National Council for Civil Liberties, *50,000 Outside the Law: An Examination of Those Certified as Mentally Defective* (London: NCCL, 1951)

Noll, Stephen, *Feeble-Minded in Our Midst: Institutions for the Mentally Retarded in the South, 1900–1940* (Chapel Hill: University of North Carolina Press, 1995)

Owen, John, *Social Darwinism and Social Policy: The Problem of The Feeble-Minded 1900–1914*, unpublished PhD dissertation (Institute of Historical Research: 1997)

Pacifist Service Units, *Problem Families, An Experiment in Social Rehabilitation* (London: Pacifist Service Units, 1946)

Penrose, Lionel, *Mental Defect* (London: Sidgwick & Jackson, 1933)

Platts-Mills, John, *Muck, Silk and Socialism: Recollections of a Left-Wing Queen's Counsel* (Wedmore: Paper Publishing, 2001)

Potts, Maggie and Fido, Rebecca, *'A Fit Person to be Removed': Personal Accounts of Life in a Mental Deficiency Institution* (Plymouth: Northcote House, 1991)

Prichard, James Cowles, *Treatise on Insanity and Other Disorders Affecting the Mind* (London: Sherwood, Gilbert and Piper, 1835)

Raphael, Winifred and Peers, Valerie, *Psychiatric Hospitals Viewed by Their Patients* (London: King Edward's Hospital Fund for London, 1972)

Report of the Seventeenth National Conference of Labour Women, 19–21 May 1936 (pamphlet)

Richardson, Angelique, *Love and Eugenics in the Late Nineteenth Century: Rational Reproduction and the New Woman* (Oxford: Clarendon, 2003)

Robertson, Gary, 'I'm Not Mad, Just Broken-Hearted', *Richmond Times-Dispatch*, 23 February 1980

Robins, H., 'Mental Deficiency and Sterilisation', *The Catholic Times*, 12 July 1929

Rolph, Sheena, *The History of Community Care for People with Learning Difficulties in Norfolk 1930–1980: The Role of Two Hostels*, unpublished PhD thesis (Open University: 1999)

Roxan, David, *Sentenced Without Cause: The Story of Peter Whitehead* (London: Frederick Muller, 1958)

Ruiz, Violeta, '"A Disease that Makes Criminals": Encephalitis Lethargica (EL) in Children, Mental Deficiency, and the 1927 Mental Deficiency Act', *Endeavour*, vol. 39 (2015), no. 1, 44–51

Sanger, Margaret, *Pivot of Civilisation* (London: Jonathan Cape, 1923)
Saunders, Janet, 'Quarantining the Weak Minded: Psychiatric Definitions of Degeneracy and the Late-Victorian Asylum', in Bynum, W.F., Shepherd, Michael, and Porter, Roy (eds.), *The Anatomy of Madness: Essays in the History of Psychiatry*, vol. 3 (London: Routledge, 1988), 273–96
A Scientific Correspondent, 'The Sterilisation of Defectives', *New Statesman*, 25 July 1931
Starkey, Pat, 'The Feckless Mother: Women, Poverty and Social Workers in Wartime and Post-War England', *Women's History Review*, vol. 9 (2000), no. 3, 539–57
Sutherland, Gillian, *Ability, Merit and Measurement: Mental Testing and English Education, 1880–1940* (Oxford: Clarendon, 1984)
Szreter, Simon, *Fertility, Class and Gender in Britain, 1860–1940* (Cambridge: Cambridge University Press, 1996)
Tabor, Mary C., 'Elementary Education', in Booth, Charles (ed.), *Life and Labour of the People in London*, Series 1, 'Poverty', vol. 3, Part 2, Ch. 2 (London: Macmillan, 1902)
Thomson, Mathew, *The Problem of Mental Deficiency: Eugenics, Democracy and Social Policy* (Oxford: Clarendon, 1998)
Tomlinson, Charles, *Families in Trouble, An Enquiry into Problem Families in Luton* (Luton: Gibbs, Bamforth & Co., 1946)
Tredgold, A.F., 'The Feeble Minded', *The Contemporary Review*, vol. 97 (1919)
Tredgold, A.F., *Mental Deficiency (Amentia)* (London: Baillière & Co., 1908)
Tredgold, A.F., *A Note on the Sterilization of Mental Defectives* (London: Central Association for Mental Welfare, 1930)
Tuke, Daniel Hack, *Dictionary of Psychological Medicine* (London: J. & A. Churchill, 1892)
Walker, Nigel and McCabe, Sarah, *Crime and Insanity in England*, vol. 2 (Edinburgh: Edinburgh University Press, 1973)
Watson, Stephen, *The Moral Imbecile: A Study of the Relations Between Penal Practice and Psychiatric Knowledge of the Habitual Offender*, unpublished PhD dissertation (University of Lancaster: 1988)
Webb, Sidney, 'The Difficulties of Individualism', Fabian Tract 69 (London: Fabian Society, 1896)
Wedgwood, C[icely] V[eronica], *The Last of the Radicals: Josiah Wedgwood MP* (London: Jonathan Cape, 1957)
Wedgwood, Josiah, *Memoirs of A Fighting Life* (London: Hutchinson, 1940)
Westwood, Louise, 'Care in the Community of the Mentally Disordered: The Case of the Guardianship Society, 1900–1939', *Social History of Medicine*, vol. 20 (2007), no. 1, 57–72
Wofinden, R.C., 'Problem Families', *Public Health*, vol. 57 (1944), 136–9

Women's Group on Public Welfare, Hygiene Committee, *Our Towns, A Close-Up: A Study Made in 1939–42, with Certain Recommendations* (London: Oxford University Press, 1943)

Zedner, Lucia, *Women, Crime and Custody in Victorian England* (Oxford: Clarendon, 1991)

Newspapers and periodicals

ABPC News Letter
Anglo-Saxon Review
Birmingham Post
British Medical Journal
Catholic Times
Civil Liberty
Common Cause
Contemporary Review
The Courier and Advertiser
Daily Herald
Daily Mail
Daily Mirror
Daily News
Eugenics Review
Evening Standard
Guardian
Hastings & St Leonard's Observer
John Bull
The Labour Leader
Lancaster Guardian & Observer
The Lancet
Leicester Mercury
Mendel Journal
The Month
Newcastle Daily Chronicle
Newcastle Evening Chronicle
News of the World
Nottingham Evening Post
The Practitioner
Reynolds's Illustrated News
Science
The Socialist
Strand

Sunday Chronicle
Sunday Pictorial
Sussex Express
The Times
Truth
The Vote
Weekly Sun
Western Mail
Westminster Review
Woman's Signal

Parliamentary Papers

Hansard, House of Commons Debates
Minutes of Evidence of the Select Committee on Lunacy Law (London: HMSO, 1877)
Report of the Royal Commission on the Penal Servitude Acts (London: HMSO, 1878–9)
Report of the Commission on Criminal Lunacy (London: HMSO, 1882)
Report and Minutes of Evidence of the Inter-Departmental Committee on Physical Deterioration (London: HMSO, 1904)
Report and Minutes of Evidence of the Royal Commission on the Care and Control of the Feeble-Minded (London: HMSO, 1908)
3 & 4 Geo. 5., ch. 28, Mental Deficiency Act (London: HMSO, 1913)
Board of Education, Report of the Consultative Committee on Psychological Tests of Educable Capacity and Their Possible Use in the Public Education System (London: HMSO, 1924)
Report of the Departmental Committee on Sexual Offences Against Young Persons (London: HMSO, 1925)
Report of the Mental Deficiency Committee ('The Wood Report') (London: HMSO, 1929)
Report of the Departmental Committee on Sterilisation ('The Brock Report') (London: HMSO, 1933)
Report of the Royal Commission on the Laws Relating to Mental Illness and Mental Deficiency ('The Percy Commission') (London: HMSO, 1957)
7 & 8 Eliz. 2., ch. 72, Mental Health Act (London: HMSO, 1959)

Archives
London Metropolitan Archives, Papers of the London County Council Mental Deficiency Sub-Committee
LCC/PH/MENT/1/3
LCC/PH/MENT/01/005
LCC/PH/MENT/01/007
LCC/PH/MENT/2/002
LCC/PH/MENT/05/002
LCC/MIN/720

National Archives
Home Office papers HO 45/10557/166505; HO 144/1088; HO 144/1098/197900
Ministry of Agriculture, Fisheries and Food MAF 258/27
Ministry of Education papers ED 50/124
Ministry of Health papers MH 51/172; MH 51/453; MH 51/454; MH 51/455; MH 51/456; MH 51/457; MH 51/712; MH 58/74; MH 58/97; MH 58/103; MH 79/291; MH 85/143; MH 85/147; MH 85/152; MH 85/166; MH 85/167; MH 86/11; MH 88/29; MH 121/44; MH 137/405
Records of the Central Criminal Court CRIM 1/2361
Records of the Lord Chancellor's Office LCO 2/3654
Treasury papers T 161/149

National Council for Civil Liberties archive, Hull History Centre
U DCL/83/4
U DCL/83/6
U DCL/83/8
U DCL/83/9

Wellcome Collection Archives
National Association for Mental Health (NAMH) Mental Deficiency Sub-Committee, 'Notes on Sterilisation, 1951', SA/MIN/A/2/6
National Council for Lunacy Reform, Minutes, Wellcome Collection Archives, SA/MIN/A/1/3

Websites
All links were correct as of 10 January 2024.
Alzheimer's Society, 'Deprivation of Liberty Safeguards (DoLS)' (www.alzheimers.org.uk/get-support/legal-financial/deprivation-liberty-safeguards-dols)

Department of Health, 'The Family Nurse Partnership Programme Information Leaflet', 2012 (https://assets.publishing.service.gov.uk/government/uploads/system/uploads/attachment_data/file/216864/The-Family-Nurse-Partnership-Programme-Information-leaflet.pdf)

Essex Chambers, 'London Borough of Hillingdon v Neary' (www.39essex.com/information-hub/case/london-borough-hillingdon-v-neary-0)

Final Report of the Commission of Investigation into Mother and Baby Homes, 12 January 2021 (www.gov.ie/en/publication/d4b3d-final-report-of-the-commission-of-investigation-into-mother-and-baby-homes/)

Gilbert, Martin, 'Leading Churchill Myths: Churchill's Campaign Against the "Feeble-Minded" Was Deliberately Omitted by His Biographers', 17 April 2013, International Churchill Society (https://winstonchurchill.org/publications/finest-hour/finest-hour-152/leading-churchill-myths-churchills-campaign-against-the-feeble-minded-was-deliberately-omitted-by-his-biographers/)

The Holocaust Encyclopedia, 'Euthanasia Program and Aktion T4' (https://encyclopedia.ushmm.org/content/en/article/euthanasia-program)

ITV News, '"Barbaric": Hundreds with Learning Disabilities Kept Locked up for Years', 2 November 2022 (www.itv.com/news/2022-11-02/barbaric-hundreds-with-learning-disabilities-kept-locked-up-for-years)

Joseph, Sir Keith, 'Speech at Edgbaston ("Our Human Stock is Threatened")', 19 October 1974, Margaret Thatcher Foundation (www.margaretthatcher.org/document/101830)

Legal Action Group, 'The Court of Protection: Steven Neary's Story' (www.lag.org.uk/article/202467/the-court-of-protection-steven-neary-s-story)

Mencap, 'No Freedom, No Dignity, No Life' (www.mencap.org.uk/get-involved/campaign-mencap/transforming-care-homes-not-hospitals)

Neary, Mark, 'Love, Belief and Balls' (https://markneary1dotcom1.wordpress.com/about/)

NHS, 'Learning Disability Services Monthly Statistics' (https://digital.nhs.uk/data-and-information/publications/statistical/learning-disability-services-statistics/at-january-2023-mhsds-november-2022-final)

OBITER J, 'Law and Lawyers: Deprivation of Liberty: the Worrying Case of Stephen Neary' (http://obiterj.blogspot.com/2010/12/deprivation-of-liberty-worrying-case-of.html)

Proceedings of the Old Bailey, 1674–1913 (oldbaileyonline.org)

Spartacus Educational (Spartacus-educational.com)

Robson, Evie and Kitzinger, Celia, 'Inspired by Bournewood: A s.21A challenge and delay in the court', post dated 10 May 2021, Open Justice Court of Protection Project – Promoting Open Justice in the Court of Protection (https://openjusticecourtofprotection.org/2021/05/09/inspired-by-bournewood-a-s-21a-challenge-and-delay-in-the-court/)

Series, Lucy, The Small Places (https://thesmallplaces.wordpress.com/)
Social Care Institute for Excellence, 'What Are Liberty Protection Safeguards?' (www.scie.org.uk/mca/lps/latest#:~:text=LPS%20will%20be%20about%20safeguarding,accommodation%20and%20their%20own%20homes)

Picture Credits

Pages 13, 15, 28 and 202: Wellcome Collection, Public Domain Mark
Pages 21 and 58: © 2006 ProQuest Information and Learning Company Document from ProQuest's House of Commons Parliamentary Papers, displayed with permission of ProQuest LLC
Page 27: Reach plc (image created courtesy of the British Library Board)
Page 42: Birmingham City Libraries (left); Lancashire County Museum Service (right)
Pages 71 and 84: Keele University Library Special Collections, Public Domain www.keele.ac.uk/library/specarc/collections/josiahclementwedgwood/
Page 76: City College of New York. Public domain
Page 90: Mencap Northern Division/David Barron/Edwin Banks
Page 95: Lancashire County Museum Service (top); Mary Evans/Peter Higginbotham Collection (bottom)
Page 97: Royal Institute of British Architects/RIBA Collections
Page 101: National Archives
Page 102: Mary Evans/Peter Higginbotham Collection
Pages 104 and 217: John Bull, successor rightsholder unknown
Page 138: Wellcome Collection, Creative Commons Attribution (Attribution 4.0 International (CC BY 4.0)
Page 143: Luton Borough Council
Page 154: Look and Learn Picture Archive
Pages 192, 251 and 252: successor rightsholders unknown
Page 201: Arthur Estabrook Papers, M.E. Grenander Special Collections & Archives, University at Albany, SUNY
Page 211: Dorset History Centre
Pages 234 and 238: Hull University Archives, Hull History Centre / Liberty

Index

Ackland, Valentine (poet) 210–3
agriculture 94, 149, 204, 224–5
Aktion T-4 programme 207
alcoholism 3, 7, 10, 11, 12, 20, 29, 46, 50, 55, 133, 136, 142, 144, 152, 159, 164, 204, 206, 266
anonymity (of case studies) xiii, 36, 114, 176, 230, 238, 267
aristocracy 5, 9, 31, 46, 194
Army 55, 94, 107, 155, 158, 162, 233, 234
Ashworth, *see* Moss Side Hospital
Assessment and Treatment Units 259, 260, 261
Attlee, Clement 145, 223, 240
autism 259–62
aversion therapy 161

baby-farming 36
Barron, David (colony inmate) 89–92, 96, 99, 129, 246
bed shortages 95–6, 168, 173–5, 179, 186, 207, 260, 273–4
Beesley, Ian (photographer) 267
Bentley, Derek 171, 172, 173
Besford Court Catholic Mental Welfare Hospital 100–2, 168
Bevan, Aneurin 125, 231
Beveridge, William 62

Binet-Simon tests, *see* intelligence/IQ
Birmingham 41, 59, 105, 117, 253, 254
birth rate, *see* fertility
blackmail 157, 160–1, 163
Blair, Tony 254–5
Board of Control 63, 93–94, 96, 98, 99, 105, 108, 109, 110, 111, 112, 113, 117, 129–132, 135, 136, 138, 139, 140, 156, 157, 158, 161, 162, 163, 166, 168–9, 185, 186, 212, 218, 225, 226, 235, 266
   abolished 244
   created 59, 64, 85
   criticised for high-handedness 137, 195, 216, 217, 229, 236–7, 238, 262
Boer Wars 55, 56, 283
borstal 12, 136, 137, 253
Bowes-Lyon, Katherine and Nerissa 156
Bradford 20, 26
Brewster, Bertie (freed patient) 216–17
Brighton Guardianship Society 111, 113, 115, 215–16, 221, 225
British Medical Association 23, 61, 188
British Medical Journal 15, 188

Broadmoor Hospital 16, 128, 167, 169–70, 183
Brock Committee on Sterilisation, 1932–4 186–7, 195, 197, 199
Buck v. Bell lawsuit 200–2
Bullying, see workplace bullying
Burt, Cyril (psychologist) 121–4, 154–5
Butler, Rab (Home Secretary and Education Secretary) 174–5

Caird, Mona (author) 47
Calderstones colony (later, hospital) 223, 249, 250
California 198–200, 206
Carroll, Lewis 108
Castle, Barbara 253
castration, see also sterilisation, 183–4, 203–4, 254
Caterham, see St Lawrence's Mental Hospital
Catholics, see Roman Catholics
Chamberlain, Neville 191
Charity Organisation Society 61
Cheshire 41
Chesterton, G.K. 67, 78, 79–80, 148–9, 150
child guidance clinics 121, 167, 244
child sexual abuse, see also incest and sex crimes 35, 45, 50, 69, 74, 115, 118, 120, 122–4, 126, 158, 164, 174–9, 196, 268, 271
Church of England 19, 48–50, 61–2
Churchill, Winston 62–7, 183, 186, 209, 242

Clouston, Dr Thomas 35
colonies for the mentally deficient, see also cruelty, labour exploitation, sex segregation, shortages of beds and staff shortages xviii, 58, 66, 68, 69, 77, 80, 92–106, 111, 114, 126, 128, 129, 131, 136, 137, 138, 150, 151, 165, 168, 172, 175, 191, 213, 218, 219, 220, 221, 223, 224, 231, 233, 236, 237, 239, 241, 245, 248, 249, 256, 257, 266
  Brockhall 95
  Calderstones 223, 249, 250
  corruption within system 102–3, 229–30
  Darenth 138, 172
  design of 92–6
  Dovenby Hall 131
  Etloe House 218
  Farmfield 136, 137, 172
  Hortham 168
  Meanwood Park 96–9, 245, 287
  numbers detained in 215
  Prudhoe Hall 94, 95, 128, 129, 175
  St Catherine's, Doncaster 104–5, 225, 226, 251
  St Joseph's, Warwickshire 100–102
  Sandhill Park 93
  Stoke Park 104, 191
  Virginia Colony at Lynchburg, US 200–1, 202
  Whixley 89–92, 96, 98, 246
community care 173, 243, 244, 245–6, 248, 252–3, 259, 260, 267

Conservative Party 48, 66, 67, 68, 71, 93, 174, 190, 195, 196, 253, 254
Contagious Diseases Acts 11–12, 120
contraception 31, 187, 196, 253, 254
Court of Protection 262–4, 278
Crichton-Browne, Sir James (doctor) xii
crime statistics 164–5, 176–9,
Criminal Law Amendment Act, 1885 39–40, 160
Crooks, Will, MP 75–6, 146, 188
cruelty to patients 69, 90, 98, 99, 100–102, 231, 234, 242, 260, 261
Culpin, Dr Millais (psychoanalyst) 133–4
cutbacks, economic 94, 111, 117, 194, 237

Daily Herald 27, 59, 79, 167
Daily Mail 45
Dangerfield, George (author) 60
Darenth Colony 138, 172
Darwin, Charles 3, 5, 33, 34, 58
deafness 22, 40, 118, 126, 127, 128, 129, 130, 132, 206
death penalty 170, 171, 172, 173
Defence of the Realm Acts, 1914 80
degeneration xviii, 3–4, 6–7, 8, 21, 29, 46, 55, 253
Dendy, Mary (educationalist) 41–2, 62, 74–5, 142
Denmark 203–4, 205
Departmental Committee on Sexual Offences Against Young Persons, 1925 175–9

Deprivation of Liberty Orders 259, 261–3, 277
Despard, Charlotte (socialist and feminist) 67
Dickens, Charles, *see also Oliver Twist* 36, 48
Dovenby Hall colony 131
dyslexia 110

Earlswood Hospital 156
East Chaldon, Dorset 210–2
Eichholz, Dr Alfred 28–30, 31, 57
Elementary Education Act, 1870 18, 19
Elementary Education (Blind and Deaf Children) Act, 1893 22
Elementary Education (Defective and Epileptic Children) Acts, 1899 and 1914 22, 266
Ellis, Havelock (sexologist) 16–17
encephalitis lethargica ('sleepy sickness') 214
environment (considered more important than inherited traits) 3, 4, 14, 17, 20, 22, 28–9, 32–3, 43, 47, 55, 57, 58, 59, 79, 115, 122, 133–4, 143, 144, 145, 148, 149, 155, 188, 214, 254
epilepsy 17, 22, 58, 116, 141, 159, 184, 200, 204, 206, 213, 233, 266
'Epping Case' 66–7
escapes 98, 101, 125, 128, 167, 169
Etloe House colony 218
eugenics/eugenicists 14, 31, 33, 34, 57, 58, 60–2, 64–5, 73, 80, 122, 127, 132, 135, 140, 141, 144, 145, 147,

155, 159, 186, 187–189, 190, 191–2, 193, 213, 233, 253, 255, 256, 277
1912 conference in London 65
seen as 'cranks' 59, 73, 191–192
Eugenics Review 33, 64, 191
Galton introduces concept 3, 8
and the left 46, 78–9, 80, 188, 196–7
hostility from medical profession 61, 184
In Nazi Germany 205–9
use of 'pedigrees' and family trees 8, 28, 141, 144, 158–9, 184, 188, 201, 214
loss of respectability as a belief 144, 207–9
persuasiveness of rhetoric xviii, 3, 59, 61
In Scandinavia 203–5
in the United States 196–203
and women 32, 43–8
Eugenics Education Society/Eugenics Society 41, 42–3, 60–2, 67, 68, 73, 122, 141, 144, 186, 195–6, 198, 199, 207, 208
Eugenics Record Office 199, 200, 201
euthanasia 191, 206–7
Evans, Timothy (miscarriage of justice victim) 171, 172, 173
evolution 5, 6, 7, 9, 16, 33
experts, distrust of 47, 72–3, 152, 167, 170, 172
exploitation, *see* labour exploitation

Fabian Society 76
Fairfield, Dr Letitia (lawyer and doctor) 123, 170–2
Farmfield State Institution 136, 137, 172
farming, *see* agriculture
Feeble-Minded Control Bill, 1911, *see* Mental Deficiency Bills
Feeble-Minded Persons (Control) Bill, 1912, *see* Mental Deficiency Bills
female sexuality, seen as problematic 34, 35, 37, 38–9, 43, 46–7, 48, 50–1, 83–4, 107, 111, 113, 116, 118–20, 122–5, 186, 203, 204, 233, 244, 257, 273–4
feminism/feminists 12, 38, 43, 45–7, 196, 203–4
fertility 5, 31–34, 112, 143, 185, 187, 188, 196, 253, 254
First World War xviii, 50, 93, 115, 121, 158, 162, 173–4, 196, 205, 213
Flügel, John (psychoanalyst) 124–5
'Forgotten Man' 239
Freedom Defence League 77
Fryd, Judy (campaigner) 151, 152
funding, *see* cutbacks, economic

Galton, Francis 1, 8, 9, 147, 149
Gateshead 127, 184–5
Geddes Axe 173
Germany 19, 55, 80, 146, 205–9
guardianship 77, 80–2, 84, 106, 111, 112, 157, 163, 164, 179, 187, 200, 213, 215–26, 266

Habitual Criminals Act, 1869 11
Haldane, J.B.S. (scientist) 188

Hardie, Keir 78, 79
head injuries 213, 262
'high-grade' patients 23, 42, 89, 92, 93, 94, 99, 103, 110, 136, 141, 168, 170, 179, 186, 193, 224, 240, 244, 273, 275
Hodson, Cora B.S. (eugenicist) 198–200, 208
Hogben, Lancelot (biologist) 155, 188–9,
Holloway Prison 66, 67, 139, 184
Home Office 23, 44, 59, 63, 64, 65, 174, 183
Homicide Act, 1957 173
homosexuality 81, 91, 123, 156–65, 215, 221, 271
hooligans/hooliganism xvii, 4
Hortham Colony 168
hostels 94, 174, 215, 220, 244, 246, 247, 253
Human Betterment Foundation 199–200
Huxley, Julian 188

Idiots Act, 1886 36, 39, 48
illegitimacy xvii, 5, 26, 35–6, 37, 38, 51, 81, 83, 102, 107, 110, 114, 115, 116, 118, 122, 125, 126, 174, 184, 219, 233, 243, 251–2, 253, 257, 270
inbreeding 5, 9, 149
incest 37, 38, 50, 115, 124, 126, 165, 176–7, 221, 271
Independent Labour Party, see Labour Party
Indiana 63, 64, 200, 274
Inebriates Act, 1898 11

institutionalised patients 114, 247, 251, 267, 268, 269, 270, 271
intelligence/IQ 15, 16, 17, 20, 22, 49, 56, 73, 112, 115, 121, 127, 134, 135, 143, 148–8, 149–55, 161, 167, 170, 173, 174, 178, 184, 200, 201, 202, 214, 218, 229, 231, 235, 253
  tests devised by Binet and Simon 149, 150, 154
  Cyril Burt on 121, 154–5
  G.K. Chesterton on 149–50
  falsification of results 155
  fears of national decline in IQ 31, 56
  people with 'normal' or high IQs detained in colonies xviii, 115, 229, 237, 241, 267
  opposition to tests from parents and teachers 151–2, 204
  Porteus Maze 150
  psychopaths and IQ 14, 15, 85, 155, 179, 243
  public view of tests 152
  rural vs urban test discrepancy 150
  and social class 155
  intelligence as a spectrum 148, 150
  word association 150
Inter-Departmental Committee on Physical Deterioration 55–6

Jackson, Mark (historian) 150
Joseph, Sir Keith 253–4

Kent 'Visitors' 135, 136
Kerr, Dr James 26, 28

Kleptomania, *see also* thieves/theft 133–4
Kropotkin, Peter 33

labour exploitation 55, 58, 101, 102, 105, 109, 113–16, 131, 146, 223–6, 231, 237, 247
Labour Party 20, 66, 75–6, 78, 79, 125, 145, 188, 194, 196–7, 223, 224, 231, 253, 254
Lancaster Moor Hospital 248–9
The Lancet 41, 133, 237, 250, 255
Lenchitsky, Zalig (murderer) 172–3, 174
Lewis, Dr E.O. 140–1, 145, 149–50
Liberal Party 59, 62, 66, 69, 70, 74, 79, 160, 196, 209
libertarianism/libertarians 12, 70, 74, 77–8, 80, 135
liberty xviii, 10–11, 65, 68, 70, 73–4, 75, 80, 153, 185, 186, 195, 232, 242, 250, 259, 261, 262, 263, 264
Liberty Protection Safeguards 264
Lloyd George, David 60, 62
Local Government Board 13, 36, 49
Lombroso, Cesare 16–17
London Association for the Care of the Mentally Defective 115–6
London County Council 22–6, 28, 38, 44, 100, 115, 116, 121, 136, 138, 154, 196
London Metropolitan Archives 25, 119, 120, 121, 165
London School Board 20, 22–3
London School of Economics 155
lumpenproletariat 62, 147

lunacy, *see* mental illness
Lunacy Act, 1890 36
Luton 141, 142–4, 146
Lynchburg, Virginia 200, 202, 275

Magdalene laundries 256
malnutrition 28, 55, 149
Manchester 37, 41, 246, 283
marriage ban 59, 68, 132, 257
Maudsley, Dr Henry 7–8
Maxwell Fyfe, David (Home Secretary) 171, 173
Mazumdar, Pauline (historian) 61, 189, 277
McKenna, Reginald (Home Secretary) 44, 68, 83, 84
Meanwood Park Colony 96–9, 245, 287
meningitis 25, 213
mental age 113, 149, 164, 168, 170, 171, 172, 200, 212, 235–6, 249, 272
Mental Capacity Act, 2005 261–2, 263
Mental Deficiency Bills and Acts
  Feeble-Minded Control Bill, 1911 62–3, 64–5
  Feeble-Minded Persons Control Bill, 1912 46–7, 48, 67–8, 188
  Mental Deficiency Bill/Act, 1913 xvii, 48, 68–9, 70–4, 75–7, 78–80, 81–5, 120, 188
  Mental Deficiency (Amendment) Act, 1927 xi, 214
  Act repealed in 1959 244
Mental Health Act, 1959 xix, 92, 244, 245, 249, 250, 252

Mental Health Act, 1983 92, 260
Mental Hospital Nurses' Association
  194
mental illness 20, 21, 22, 23, 36, 49,
  51, 54, 61, 86, 103–4, 114, 128,
  161, 170, 177, 178, 185, 194–5,
  199, 200, 203, 205, 206, 207, 242,
  247–8
Mercier, Dr Charles 14, 57
meritocracy 46
Metropolitan Association for
  Befriending Young Servants 39,
  115
Mid Yorkshire Institute for the
  Mentally Defective, *see* Whixley
  Colony
Millbank Penitentiary 10
M'Naghten Rules 170–1
moral imbecility/moral defectiveness
  xvii–xviii, 23–4, 25, 44–5, 50,
  80, 102–3, 110, 113, 114, 117,
  118, 119, 121, 125, 126, 127,
  128, 129, 130, 133, 136, 154,
  157, 161, 163, 165, 179, 186,
  201, 204, 219, 223, 239, 248,
  249, 251, 257, 267, 268
  change in terminology 1927 xi, 214
  definition xii
  uncertain meaning xvii, 14–16,
    84–5, 109, 132, 135, 242–4
moral insanity 6, 7, 8, 14, 133
Morel, Bénédict-Augustin
  (psychiatrist) 6–7, 8
Morrison, Herbert 196
Moss Side Hospital 100
mothers/motherhood 32, 36–7, 43–4,
  69, 83, 109, 114, 124, 125, 126,
  144, 145, 146, 151–2, 243, 253,
  254, 256, 270, 275

National Association for the Care and
  Protection of the Feeble-Minded
  41–2, 61, 62, 68, 74
National Association of Parents of
  Backward Children (Mencap)
  151–2
National Council for Civil Liberties
  xvii, 226, 229–30, 232–41, 242,
  243, 250, 257, 277
National Council for Lunacy Reform
  194–5
National Insurance 60, 240
National Vigilance Association 37,
  124, 161
National Workers' Committee for the
  Legalising of Voluntary
  Sterilisation 196
Nazism 144, 146, 205–9
Neary, Steven, and Neary, Mark 260–1
neglect, of children and youngsters
  20, 36, 50, 66, 69, 74, 82, 107,
  124, 125, 144, 146, 236, 239,
  240–1, 243, 265
Newcastle 73, 94, 101, 174
Newsholme, Arthur (medical officer/
  public health expert) 33
newspaper press 12, 27, 32, 59, 73, 79,
  100, 101, 103–4, 105, 111, 128,
  141, 167, 170, 174, 176, 190, 192,
  193, 194, 208, 216–19, 229,
  232–3, 239, 247, 250–1, 254, 257,
  261

NHS 174, 231, 254, 260, 261, 271,
nineteenth century, *see* Victorian era
NSPCC 37–8, 39–40, 66, 74, 75
Nurse Family Partnership 254–5

Oliver Twist 36
oral history 92, 96, 245, 247–8
Oxford 4, 107, 108–9, 110, 111, 112

patient confidentiality, *see* anonymity
Pearson, Karl (biostatistician) 31, 56, 150
Penrose, Lionel (psychiatrist and geneticist) 147–8
Percy Commission, *see* the Royal Commission on the Laws Relating to Mental Illness and Mental Deficiency, 1957
Percy, Eustace, Lord 190–1
permanent detention 10, 26, 41, 42, 43, 44, 58, 62, 64, 66, 68, 72, 74, 75, 85, 108, 110, 123, 127, 148, 152, 270
Personal Rights Association 11
Pinsent, Ellen (educationalist) 41, 42, 62
Platts-Mills, John, QC 102–3, 240, 241
police 11, 65, 91, 98, 107, 117, 126, 128, 132, 139, 140, 160, 161, 164, 168, 169, 170, 171, 172, 216, 222, 229, 235
Poor Law authorities 29, 32, 35, 36–7, 39, 47, 49, 50, 56, 60, 67, 77, 94, 107, 109–10, 117, 120, 140, 141, 243

poverty 4, 9, 34, 36, 66, 76, 78, 79, 84, 136, 140–7, 253
Prevention of Crimes Act (1871) 11 and (1908) 11, 65
preventive detention xviii, 11, 65, 239
Prichard, James Cowles (evolution theorist) 6
prison 16, 19, 25–32, 46, 53, 61, 62, 76, 78, 81, 88, 120, 237, 253
  Holloway 66–7, 139, 184
  Oxford 107
  in United States 200
Prison Act, 1898 16
prostitution/prostitutes, *see also* Contagious Diseases Acts 10, 11–12, 35, 43, 44, 45, 49, 50, 111, 119–23, 147, 204
Prudhoe Hall Colony 94, 95, 128, 129, 175
psychoanalysis 79, 124, 133, 147, 150–1
psychology xviii, 50, 72, 102, 120, 121–2, 147–8, 150, 153, 154, 155, 172, 176, 214, 243, 257
psychopathy/psychopaths 14–15, 17, 85, 151, 170–1, 243–4
public opinion 12, 59, 62, 111, 146, 152, 179, 183, 194
  alleged apathy 232–3
  on crime and punishment 167, 169–71, 172, 179
  on IQ testing 132
  opposition to the mental deficiency system 60, 79, 195, 247–8, 256–7

on sterilisation 61, 185–6, 208, 209
Punishment of Incest Act, 1908 38, 176

racism 203, 205, 207–8
Radnor Commission, *see* Royal Commission on the Care and Control of the Feeble Minded
Rampton 91, 98, 128–9, 130–1, 136, 152, 170, 185, 223, 246, 250, 251, 261
rape, *see* sex crimes
recidivism 4, 10, 13, 14, 120, 178–9, 200, 239
Reisz, Karel 172
'rescue' societies (for girls) 5, 35, 39, 115, 117
Riddell, George, Lord (newspaper magnate) 190–1
Rillington Place 171
Rolph, Sheena 220, 245, 247, 277
Roman Catholics/Roman Catholicism 79–80, 94, 100–1, 188, 193, 196, 205, 208, 218
Royal Commission on the Care and Control of the Feeble Minded (the Radnor Commission), 1908 56, 57–9, 61, 63, 64, 66, 68–9, 133
Royal Commission on the Laws Relating to Mental Illness and Mental Deficiency (the Percy Commission), 1957 242–4, 245
Royal Commission on the Poor Laws, 1909 60

Royal Eastern Counties Institution 147, 216, 240
Royal Western Counties Institution, *see* Starcross Hospital
rural conditions 46, 56, 81, 132, 145–6, 148–50
Rutty, Kathleen (freed patient) 139–142

St Catherine's Mental Deficiency Colony/Institution/Hospital 104–5, 225, 226, 251
St Joseph's School, Sambourne, Warwickshire 100–2
St Lawrence's Mental Hospital, Caterham 235, 245
Salvation Army 112, 118, 270
Sandhill Park Colony 94
Sayer, Dr Ettie 33, 44–5
schools/schooling, *see also* special schools 5, 19, 28–30, 34, 47, 57, 77, 104, 121, 140, 147, 150, 151, 155, 218, 229, 235, 237, 238, 239, 243
  Assistant Schoolmasters' conference 151
  Bradford School Board 20, 26
  free school meals 20
  industrial schools 12, 35, 39, 41, 42, 48, 266
  London School Board (later LCC schools department) 20, 22–3, 24, 25, 29
  reform schools xviii, 12, 35, 41, 42, 266
  schoolchildren in poverty 20–1, 28
Scotland 38, 124, 132

Second World War 105, 147, 152, 155, 173–4, 206, 207, 219, 223, 249
Sensation Fiction 42
servants 39, 50–1, 102, 108, 109, 115, 116, 119, 123, 157, 210, 212, 226, 267
sex crimes, *see also* incest and child sexual abuse 5, 35, 38, 39, 42, 43, 50, 91, 115, 118, 122, 123, 124, 158, 160, 163–5, 166, 168, 169, 172, 173, 174–8, 183, 184, 185, 196, 200, 203, 221–2, 223, 235, 236, 249–50, 268, 271
sex segregation xviii, 41, 58, 68, 92, 96, 98, 186, 194, 204
shame xviii–xix, 51, 98, 114, 116, 126, 127, 160, 269, 271
shortage of beds, *see* bed shortages
shortage of staff, *see* staff shortages
slums 20–21, 34, 136, 145, 146, 149, 152, 155, 156
Socialism/Socialists 46, 62, 76, 77–8, 79, 195, 205, 224
Socialist Health Association 224
'social problem group' 140–7, 233, 253, 254–5
social purity movement 38, 43–4, 47
social workers 135, 144, 216, 221, 243, 245, 257, 263
Society of Labour Lawyers 125, 224
sociology/sociologists 73–4, 140, 147
Somerset 94, 117, 221
Sonoma State Hospital, California 198–9, 275
special schools 22, 24, 25, 28, 41, 57, 60, 68, 92, 148, 150, 152, 218, 221, 230, 235, 266

Spencer, Herbert (polymath) 8
staff shortages 95–6, 105, 226, 252, 259, 260, 263
Starcross Hospital 99–100, 114–5, 247, 288–9
sterilisation 26, 57, 59, 63–6, 78, 180, 183–97, 189–209, 233, 257, 272–5
Stoke Park Mental Deficiency Colony 104, 191
Straffen, John (murderer) 167–72, 174
suffrage movement 44, 46–7
Sutherland, Dr John 38
Sweden 204–5
syphilis, *see* venereal disease

Thatcher, Margaret 254
thieves/theft xvii, 10, 79, 97, 114, 126, 127, 133, 134, 136, 137, 138–9, 147, 165, 168, 186, 219, 223, 230, 235, 239, 242, 249
Thomas, Dr Rees 244
Thorpe, Dora (freed patient) 103–4, 111
Townsend Warner, Sylvia 210–2
trade unions 75, 76, 79, 105, 146, 188, 194, 196, 223–4, 225
Tredgold, Dr Alfred xviii, 33–4, 64, 83, 85
Tuke, Dr Daniel Hack 7, 8

United States of America 68, 124, 150, 151, 155, 163, 190, 196, 254–5, 270
  survey of US states' sterilisation programmes 272–5
  California 198–200, 206, 273, 274, 275, 276

Indiana 63, 64, 200, 274
Virginia 200–3, 206, 274, 275
Washington state 200, 272
Urania Cottage 48

vagrants/vagrancy 10, 107, 114, 116, 118, 165, 166, 178, 204, 207
vasectomy, *see also* sterilisation 63, 183, 187, 196
venereal disease 11, 12, 28, 32, 43, 44, 45, 55, 107, 108, 117, 119, 120–1, 125, 126, 177
Victorian era xix, 45, 51, 62, 113, 129, 145, 176, 218, 277
  mental health system 5, 6, 14, 23, 42, 69, 72, 85, 93, 96, 99, 161, 260, 266, 269
  penal system 12, 14
  women's and girls' position during 11, 32, 43–4, 48, 49–50, 115, 122, 258
voluntary associations 44, 82, 107, 111, 115, 117, 118, 212 (Dorset), 214–15, 221, 224, 226

Ward, Dame Irene, MP 174–5
Wedgwood, Josiah 70–8, 80, 83, 84, 119, 120, 185, 188, 214, 230

Wells, H.G. 32, 73, 77
West Lancashire Association for the Care of the Mentally Defective 118–19
Whitehead, Peter (freed patient) 100–101, 168
Whixley Colony 89–92, 96, 98, 246
Wood Committee on Mental Deficiency, 1929 103, 140–2, 148–9
Woolmore, Annie, *see* Epping Case
workhouse xviii, 10, 32, 35–6, 37, 41, 42, 48, 59, 68, 74, 75, 83, 94, 113
  Abingdon 107–8, 109–10
  Kensington 116
  Marylebone 116
  Ongar 66
  Paddington 116
Workhouse Girls' Aid Society 50–1
workplace bullying (of people out on licence) 220–1, 247
World War I, *see* First World War
World War II, *see* Second World War

x-rays 65, 185, 187

Yorkshire 26, 36, 89, 90, 233, 270